\

Fiber Science

FIBER SCIENCE

STEVEN B. WARNER

University of Massachusetts, Dartmouth

PRENTICE HALL, Englewood Cliffs, NJ 07632

Library of Congress Cataloging-in-Publication Data

Warner, Steven B.
  Fiber science / Steven B. Warner.
      p. cm.
  Includes index.
  ISBN 0-02-424541-0
  1. Textile fibers, Synthetic.   I. Title.
TS1548.5.W39   1995
677'.4–dc20

94-13259
    CIP

Editor: Bill Stenquist
Production Supervisor: John Travis and Helen Wallace
Production Manager: Francesca Drago
Interior and Cover Design: Robert Freese

©1995 by Prentice-Hall, Inc.
A Division of Simon & Schuster, Inc.
Englewood Cliffs, New Jersey 07632

Printed in the United States of America

10 9 8 7 6 5 4 3 2 1

0-02-424541-0

Prentice-Hall International (UK) limited, *London*
Prentice-Hall of Australia Pty. Limited, *Sydney*
Prentice-Hall Canada Inc., *Toronto*
Prentice-Hall Hispanoamericana, S.A., *Mexico*
Prentice-Hall of India Private Limited, *New Delhi*
Prentice-Hall of Japan, Inc., *Singapore*
Editora Prentice-Hall do Brasil, Ltda., *Rio de Janeiro*

PREFACE

This book was developed for the students at Georgia Institute of Technology who are studying textile and fiber engineering. It was designed to be a resource for sophomores and juniors enrolled in Fiber Science, a required course. This course is a materials course that focuses on the structure and properties of polymeric fibers. Hence, the text treats chiefly organic fibers. The emphasis is on fibers used as textile and industrial fibers. Fibers for composite reinforcement and other specialty applications are also discussed as appropriate, but the treatment of these fibers is not intended to be comprehensive.

The information in this text covers the fundamental principles of fiber science. The focus centers on the underlying principles of important phenomena and properties of fibers rather than on trying to cover all aspects of all fibers. The material is presented in a manner that emphasizes concept understanding rather than one requiring extensive memorization by the student.

The text assumes the students have taken at least three quarters of calculus, two quarters of chemistry, and one quarter of physics by the time they enroll in this course. Still, the material in the text is an ambitious undertaking for sophomores in a quarter system and is better covered in a full semester. With proper encouragement and enthusiasm, however, it is possible to give students at the sophomore or junior level a strong base in fiber science with a single course. Several simple derivations are presented in the text. They are included because they bring one or more important concepts to light. You may skip the derivations and go directly to the physical basis of the result, if you chose, without loss of continuity in the text.

Fiber Science is divided into four parts. Part I introduces fundamental concepts associated with organic fiber chemistry, the organization of molecules in fibers or microstructure and macrostructure, the importance of staple fiber length and cross-section, and environmental effects such as radiation damage and interactions of fluids with fibers. Part II is dedicated to presenting the most important aspects of mechanical properties. Tensile, compressive, torsion, and bending properties of fibers are discussed. Fiber strength and modulus are singled out as examples of extrinsic and intrinsic properties. Coverage is also afforded to measurement of fiber properties, both off-line and on-line. Real world aspects of time-dependent mechanical properties are introduced. Modeling of structure to predict or rationalize mechanical properties is broached. Part III focuses on physical properties of fibers—optical, thermal, electrical, and frictional aspects are covered. Part IV discusses properties of fabrics, covering how fiber properties directly affect the bending and wetting behavior of fabrics.

The author gratefully acknowledges the assistance provided by the following coworkers: Charles Hodges, Jay Gallman, Chuck Carr, Abhiraman, Satish Kumar, Youjiang Wang, Terry Timmons, Mary Lynn Realff, and Carl Rippl. Kimberly–Clark Corp. and Hoechst–Celanese provided me with twelve years of valuable industrial experience alongside a number of fine scientists and engineers, from whom I have learned a great deal. The Georgia Tech Foundation provided a small grant to get this text underway. The National Textile Center indirectly provided important resources without which this book would not have been possible. I am especially appreciative of my wife, Ruth, and children, Genoa, Brea, and Cali, for the support they have provided, including tolerating my absence while I wrote this book and trained for triathlons.

S. B. W.

CONTENTS

6. Environmental Effects: Solvents, Moisture, and Radiation   99

PART TWO: Mechanical Properties

7. Tensile Properties   123

8. Mechanical Properties of Fibers: Shear, Bending, Torsion, and Compression   149

PART THREE: Physical Properties

12. Optical Properties   213

13. Thermal Properties   230

14. Fiber Friction, Electrical Conductivity, and Static Charge Effects   246

PART FOUR: Fibers to Fabrics

1

Introduction to Fiber Science

Fibers have been used for thousands of years, and their importance in the world today is perhaps greater now than ever before. Look around you, and you will see fibers used in a plethora of applications:

Apparel—The clothes you wear are textile materials, made from textile grade fibers. Most clothes are woven or knit products, but some, such as felts, may be nonwoven materials. The apparel industry is roughly one third of the textile industry.

Home furnishings—Many home furnishings are all or part textile fiber. Carpets are constructed from both woven and nonwoven materials. Sofas are usually covered with heavy-duty woven fabrics. Home furnishings roughly comprise one third of the textile industry.

High-performance fibers and nonwoven materials—High-performance fibers may be used in high-performance fabrics, such as flight spacesuits or firefighters' uniforms, where the fabrics shield the wearer from high temperature. Industrial fibers are used as the reinforcement elements in flexible composite materials, such as in automobile tires or belts, as well as rigid composite materials, such as boat hulls, tennis rackets, skis, aerospace applications, and bicycle frames. Nonwoven materials are used in a number of applications, notably nondurable or disposable applications such as operating room gowns, where the materials provide cleanliness and barrier properties superior to those of woven materials. The nonwovens industry is one of the fastest-growing segments of the textile industry. This grouping of fibers comprises the remaining third of the fiber industry.

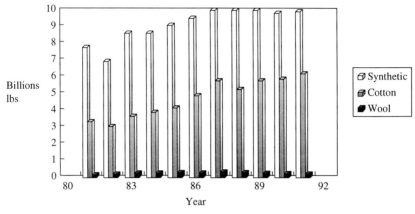

FIGURE 1.1   Mill consumption of fiber in the United States. (*Fiber Organon*, March 1991.)

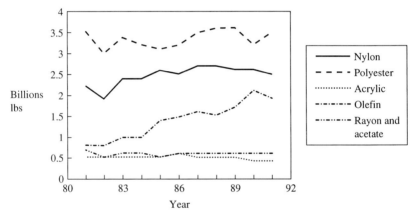

FIGURE 1.2   Mill consumption of synthetic fiber in the United States. (*Fiber Organon*, March 1991.)

All in all, about 12 percent of the jobs in the United States are in the textile field—more than 2,000,000 jobs nationwide. Textiles is the second largest industry in the United States. Americans spend nearly $202,000,000,000 on clothing and textiles per year (data from T. Malone, President of Milliken Corp., Feb. 1993, Auburn University). Figure 1.1 summarizes the mill consumption of wool, cotton, and synthetic fibers in the United States. The consumption of silk is roughly 1 percent that of wool. A breakdown of the consumption of synthetic fiber is provided in Figure 1.2.

Cotton consumption increased 8 percent, wool 22 percent, and polyester 9 percent in 1991 from the previous year. Polyester consumption remains large and growing. Polyolefin consumption is growing very quickly, in part due to improved polymers at reduced prices. The charts you have just examined do not contain data on reinforcement or specialty fibers. Of these, glass fiber boasts the largest volume, which was 1.3 billion pounds in 1984. Sales of specialty fibers are less than those for commodity textile fibers; however their price is usually high. Consider carbon fibers, for example; sales volume is roughly 1.5 million pounds, but fiber cost is on the order of $50 per pound.

Yet despite the age and maturity of fiber science, a great deal of excitement is present. New processes, materials, and products are being developed daily. Let us spend a few minutes to describe just a few of the recent advances in fiber science:

1. *Polybenzimidazoles* are heat-stable polymers that were developed and commercialized about a decade ago. PBI® fabrics are excellent for flight suits and firefighters' apparel, since the fibers are comfortable, nonburning, and stable to temperatures perhaps in excess of 600 °C.

2. The first *liquid crystal polymers* were developed about 20 years ago, but new ones are continually being developed. The most well-known LCP is Kevlar®, which has mechanical properties that are superior to those of steel on a pound-for-pound basis. These and other fibers with high mechanical properties are used in ballistic applications, such as bulletproof vests, and in composites.

3. Synthetic fibers with unusual cross-sectional shapes have been developed. *Hollow fibers* offer good resiliency and find use in insulation or high-loft nonwoven materials; multilobal fibers offer unique optical properties, such as the ability to hide dirt, and find use in carpet fiber technology; and bicomponent fibers offer the advantages of both components and find use in nonwoven materials. An example of a hollow fiber is shown in Figure 1.3.

4. Very small-diameter fibers, *microfibers*, have been brought to commercial reality in only the past several years. They offer extreme softness and high comfort.

5. Elastic materials that can be formed into fiber were developed in only the past 20 years. The most common *thermoplastic elastomer* is Lycra®, which is used in a number of stretchy apparel fabrics.

6. Polymeric fibers with *high electrical conductivity* are currently under study. Samples with conductivity nearly as high as that of copper on a weight basis have been developed, yet much more work is required before these materials appear on the marketplace.

7. Manufacturing technology has also been updated. Fiber extrusion is highly automated. Some recently upgraded fiber formation plants operate completely in the dark, since no human operator is required. Product quality may be checked in real time. New manufacturing techniques are coming onstream all the time. Two such techniques developed in the last 20 years are *meltblowing* and *spunbonding*. Both of these techniques use directed air to reduce the fiber diameter just after fiber extrusion. They are capable of processing rates that were previously unattainable.

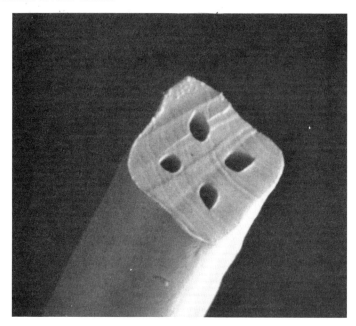

FIGURE 1.3 Cross-section of a hollow polyester fiber.

8. *Superabsorbent fibers* have been developed in the last decade. These materials absorb 100 times or more their weight in water and enabled the development of ultrathin diapers.

Examples of how new technology in fiber science has changed the world we live in are many. One example is the roof of the new dome in Atlanta, the largest rigid-cable-supported structure in the world, shown in Figure 1.4. The entire roof, Teflon®-coated woven fiberglass, is cable supported. It is composed of fiber- and fabric-reinforced composite materials. There are 20.9 acres, or 67 tons, of fabric; 11.1 miles of cables ranging from 1 to 4 inches in diameter; and 2,500 linear feet of compression ring. The structure, designed by Mathew Levy, is elegant in its simplicity. All of the stress is completely on the cables and the compression ring, rather than on the fabric. There are no posts or columns to block spectators' views. The strength of the design is shown by the capability of the roof to support an automobile.

The purpose of this text is to provide you with a basic understanding of fiber science. The focus of the text is on understanding the structure and properties of fibers. Textile fibers assume a dominant position in this text, but the structure and properties of high-performance fibers are treated to a limited extent. By structure I mean both chemical architecture and the arrangement of the molecules in the fiber or microstructure. Microstructure is considered on several scales, from the scale of atomic units, or Angstroms (10^{-10} m), to microns (10^{-6} m), all the way to about millimeter scale (10^{-3} m). Fiber processing is not emphasized, but it is treated on a need-to-know basis. The

FIGURE 1.4 Roof and support structure of Georgia dome. (Photo by Robert Reck, courtesy of Birdair, Inc.)

triangle in Figure 1.5 emphasizes the relationships between structure, properties, and processing of fibers.

Both natural and synthetic fibers will be treated in this text. Historically, natural fibers were the first textile and industrial fibers. The natural fiber used

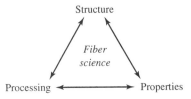

FIGURE 1.5 Diagram showing the interdependence of structure, properties, and processing of fibers.

in the highest volumes is cotton. Other plant fibers are flax, hemp, linen, etc. Wool, silk, vicuna, collagen, and spider silk are examples of animal fibers, the former being the most commercially significant. The use of natural fibers for garments dates back to at least 10,000 B.C., where flax was used in Switzerland. Cotton dates back to only 3000 B.C. Fiber science is rich in history. On the other hand, synthetic fiber formation is relatively new. Viscose rayon, what we may call semisynthetic, was made by Hilaire de Chardonnet in 1889. Viscose rayon fibers are made by modifying cellulose, a natural polymer occurring in plants; forming the fiber; and then converting the material back to cellulose. The first U.S. rayon plant was constructed in 1910. In the 1930s, polyester and nylons were developed, marking the beginning of true synthetic fibers. Since that time, synthetic fibers have assumed an increasingly important role as textile materials.

Except for silk and flax, natural fibers are relatively short staple fibers. That is, they are discontinuous fibers. The average cotton fiber is roughly 3 cm, and the average wool fiber is roughly 10 cm long. Synthetic fibers are made on a continuous basis. Synthetic industrial fibers are generally not cut, but many synthetic fibers used in textile applications are cut to staple lengths. This text does not treat to a significant extent very short fibers, such as wood pulp fibers, that are used in paper making.

Before an understanding of the structure and properties of fibers can be developed, it is critical to understand the basic chemical structure of polymers. After all, organic fibers are simply a specific shape of polymeric materials. Therefore, Chapter 2 is dedicated to teaching the chemical structure of polymers. Chapter 3 centers on the physical structure of fibers and the nature of chain aggregation or microstructure of polymers in general. The focus of both Chapters 2 and 3 is organic polymers, as most textile fibers are organic. Ceramic fibers and metal fibers are not emphasized in this text, although, because of their importance, oxide glass fibers are discussed.

The focus of Chapter 4 is on obtaining a random sample and, hence, data that reflect the mean properties of a large sample. Especially important is the fiber length distribution of natural fibers. Chapter 5 deals with the importance of fiber cross-sections and how to measure linear density in both individual fibers and large collections of fibers.

The feature topic in Chapter 6 is the effect of fiber exposure to good and poor solvents. Of special interest is moisture, since all natural and some synthetic fibers interact with water, leading to changes in properties. (Chapter

16 is a follow-up, in which the fundamental interactions of fluids with fibrous assemblies—wetting, capillary forces, and flow—are discussed.)

Part II of the text deals with mechanical properties. Tensile properties are covered in Chapter 7. This is followed by a chapter dealing with other modes of deformation—bending, torsion, compression, and shear. Chapter 9 brings up the concept of fiber-to-fiber and along-fiber variability. The effects of fiber variability on fiber and yarn properties, especially tenacity, are discussed, and so too is the value of continuous testing of fibers and yarns. Chapter 10 deals with the time and temperature dependence of (mechanical) properties—viscoelasticity. Chapter 11 closes out the section on mechanical properties. The chapter presents some basic concepts on using models of structure to understand and predict mechanical and other properties.

The next three chapters deal with important physical properties—optical properties, thermal properties, friction, electrical conductivity, and static charge buildup. A student who already is familiar with basic material properties will be more prepared for this section, but such a background is not essential.

The text on fiber science is essentially complete at this point. The final chapters show how the properties of fibers are translated into woven or knit fabrics via yarns and directly into nonwoven fabrics. These chapters are certainly not intended to be comprehensive, as there are entire books dedicated to the subjects; rather, they are designed to show how the information in the text may be extended to understand some of the properties of fabrics.

PART ONE

Fiber Fundamentals

2

Chemical Structure of Polymer Fibers

An understanding of the chemical structure of fibers is critical to an appreciation of their optical, thermal, mechanical, electrical and dielectric, and environmental properties. Microstructure is just as important as chemical structure in influencing the properties of a polymer. Hence, Chapter 2 is dedicated to chemical structure, and the chapter that immediately follows deals with microstructure. Chemical structure does not depend on the shape of the macroscopic article—be it film, fiber, or bulk—or on how the material is processed. Microstructure, on the other hand, does depend on the shape, because the required shape in part determines the appropriate processing technique. For natural polymers we will consider the macroscopic fiber itself and its structure. The chemical and microstructural variations from species to species will be largely ignored. For synthetic polymers there are many possible microstructures, even for a single polymer in fiber form. We will focus on typical structures of textile and industrial fibers.

2.1 Chemical Structure of Synthetic Fibers

Polymers are macromolecules. The word *polymer* literally means "many mers." In its simplest form, a polymer consists of a single basic chemical structural unit. In a polymer there are many such units, all identical, that are linked together with a specific type of bond. The generic name of polymers that you are

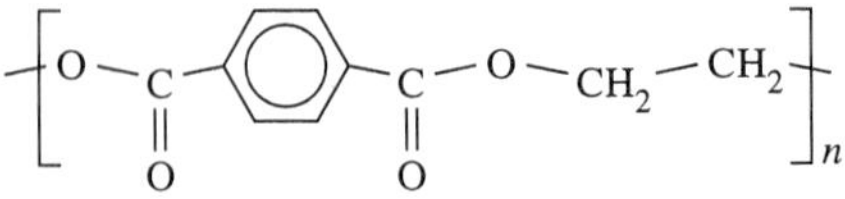

FIGURE 2.1 Chemical structure of poly(ethylene terephthalate).

familiar with describes the link connecting the repeat unit. For example, a polyester consists of units connected as shown:

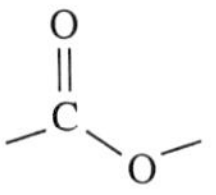

The chemistry of the material between the ester units gives the polymer its specific name. There is virtually a limitless number of ways that monomers can be combined to give polymers. We shall not investigate all the possibilities; rather, we will focus on only some of the polymers that are commercially available.

The most common polyester is PET, or poly(ethylene terephthalate). The repeat unit is in parentheses. The structure of PET is shown in Figure 2.1, where n is the number of repeat units in the molecule, i.e., the degree of polymerization. The molecular weight of the molecule can be calculated by multiplying n times the molecular weight of the mer. In reality, a fiber consists of many polymer molecules. The molecular weight is an important characteristic of the material, because it influences mechanical and other physical properties. The molecular weight of a particular sample of either natural or synthetic polymer is not monodisperse; rather, it is an average of the molecular weight of all the chains in the sample. It is indeed unlikely that any two chains would have precisely the same molecular weight. PET fibers typically have a molecular weight of 15,000 to 25,000 g/mole. The low-molecular-weight fiber is used in textile applications, and the high-molecular-weight polymer is used to make industrial yarns.

Another common polyester fiber is PBT, or poly(butylene terephthalate). Its structure is similar to that of PET except that it contains four CH_2 rather than the two in PET. This minor change in chemical structure, as we shall see, gives rise to major changes in crystal structure and, hence, to major changes in the physical properties of the fiber.

The last polyester that we will discuss is one from the rather new class of polymers called liquid crystalline polymers, or LCP for short. These polymers have excellent mechanical and other physical properties, so their applications are in industrial areas. A characteristic of liquid crystalline polymers is that the polymer backbone is rigid and, hence, usually contains aromatic groups. The structure of one common LC copolyester, Vectra®, or Vectran®, in fiber

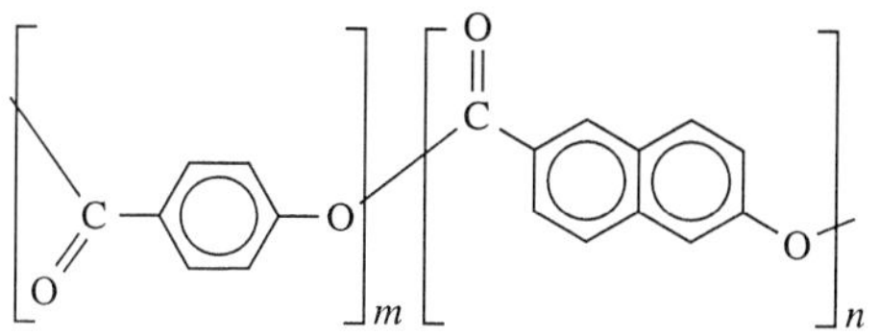

FIGURE 2.2   Chemical structure of Vectra® copolyester.

form, is shown in Figure 2.2, where *m* and *n* are the number of repeat units in each segment.

Another generic class, of polymers is nylons, or polyamides. The repeat unit in a polyamide is shown:

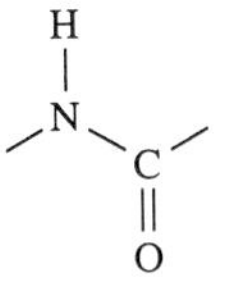

The most common polyamides are nylon 6, nylon 66, poly(meta-phenylene isophthalamide) or PMIA, and poly(para-phenylene terephthalamide) or PPTA; the latter two are commonly known as Nomex® and Kevlar®. The structure of these materials is shown in Figure 2.3. Nylon 6 and 66 are textile or industrial fibers and Kevlar®, also a liquid crystalline polymer, is an industrial fiber. Nomex® and Kevlar® are known for their good thermal stability.

Unlike the other polymers we have mentioned up to this point, PPTA and PMIA do not melt when heated; rather, they degrade. Such polymers are called thermoset, whereas those that melt are called thermoplastic. Examples of other thermoset polymers are the natural fibers—cotton, silk, wool. These polymers also degrade prior to melting. We will probe the reason for this behavior later in the chapter.

Another synthetic polymer commonly used in fiber form is PAN, or poly-(acrylonitrile). The structure of PAN is shown in Figure 2.4. Homopolymer PAN is generally not used as a textile fiber, but, rather, PAN containing a few mole percentage of a copolymer is used. The comonomers often used are methyl acrylate or vinyl acetate. Incorporation of the comonomer makes the PAN easier to synthesize, process, and dye. These copolymers may be referred to as acrylics or loosely (incorrectly) called PAN.

PAN is an example of the class of polymers called vinyl polymers. Vinyl polymers have a two-carbon, three-hydrogen atom backbone with one substituent in place of the fourth hydrogen required for complete saturation. We will see several other examples of vinyl polymers throughout this text.

Another class of important synthetic polymers is polyolefins. They are important textile and industrial fibers. These polymers contain only carbon atoms

Nylon 6

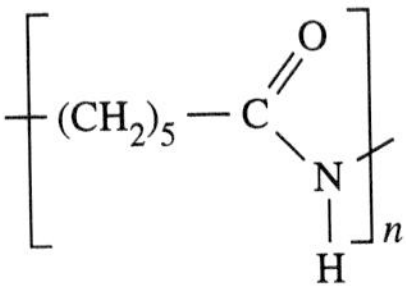

Nylon 66

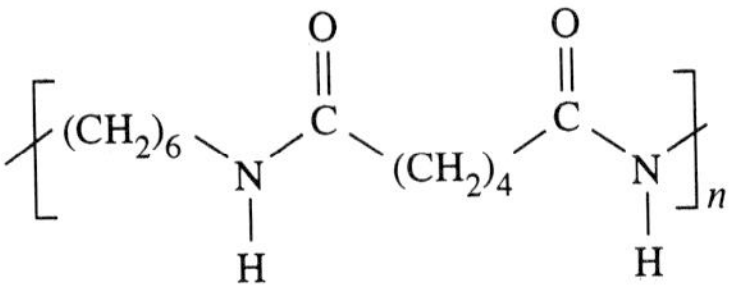

PPTA (Kevlar®)

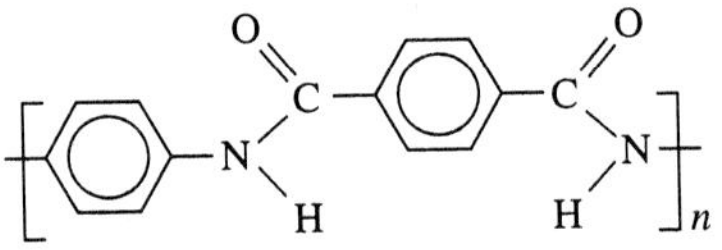

PMIA (Nomex®)

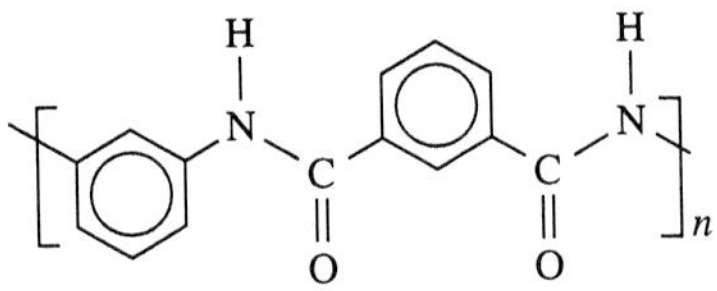

FIGURE 2.3   Chemical structure of polyamides.

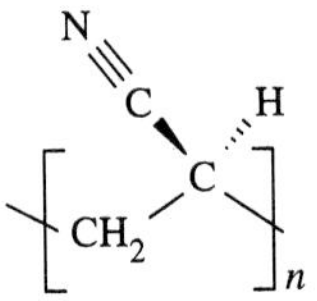

FIGURE 2.4   Chemical structure of polyacylonitrile.

in the main chain. Examples include

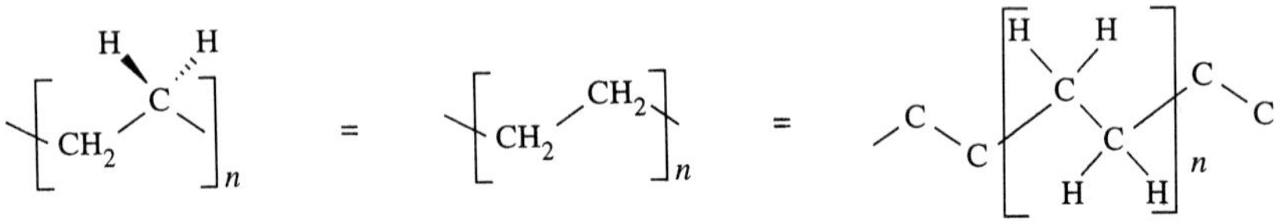

polyethylene

and

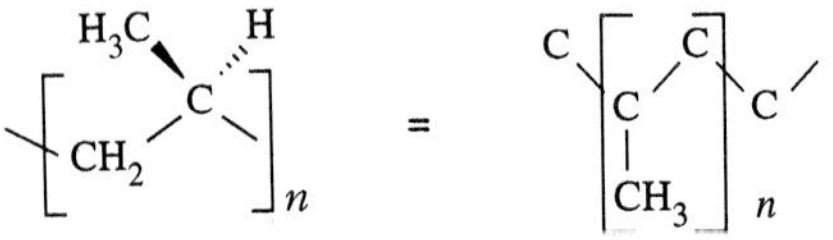

i-polypropylene

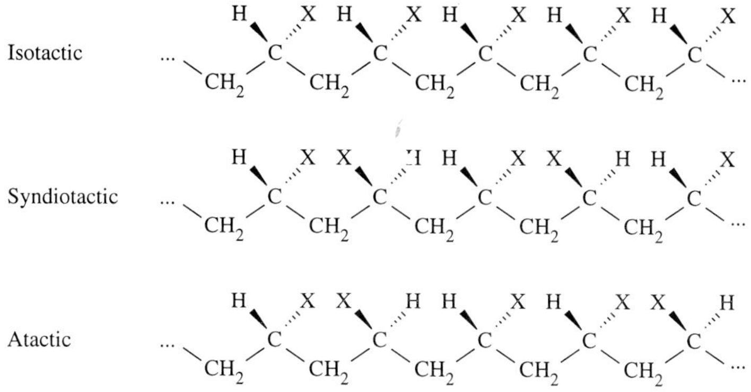

Isotactic

Syndiotactic

Atactic

FIGURE 2.5 Polymer tacticity.

The i, or isotactic, refers to the fact that the methyl groups in PP are placed in the same positions in all the mers. These positions are established during synthesis or polymerization. The other two possible side-group positions are atactic (a) or syndiotactic (s), as shown in Figure 2.5. Atactic refers to a random positioning of the side groups; isotactic, to all side groups on the same side of the polymer chain; and syndiotactic, to alternating sides. Atactic-PP is a virtually useless gummy material. The birth of the modern PP industry came with the advent of special catalysts to direct the positioning of side groups into the isotactic positions.

The positioning of the side groups is called polymer configuration. Vinyl polymers may have isotactic, syndiotactic, or atactic configuration. Conformation, on the other hand, is a term that is associated with the rotational position of atoms relative to one another. Consider a few segments of polyethylene, for example, as shown in Figure 2.6. In Figure 2.6a, the carbon atoms numbered 1 and 4 are as close together as the bond length and angle allow. This is the cis conformation. In Figure 2.6b, the 1 and 4 carbon atoms are as far apart as possible, which is the trans conformation. Another conformation with which you are all familiar is called gauche; it is illustrated in Figure 2.6c. In the gauche conformation, the polymer backbone is not coplanar; rather, the bond joining carbon atoms 3 and 4 deviates $\pm60°$ from the plane defined by atoms 1, 2, and 3. A chain that is all-trans is planar and has its ends separated as much as possible. A chain coiled on itself must contain a number of gauche or cis bonds and occupy three dimensions.

Polyethylene melts to a viscous liquid at temperatures not too much higher than those reached in commercial dryers. Hence, it cannot be used in applications requiring normal laundering. It can be and is used in disposable nonwovens. In addition, the past 10 years have seen fiber physicists develop unique ways to process PE into industrial fibers with appropriate microstructures to give extremely attractive physical properties—high strength and stiffness. These fibers are used in composite materials and other specialty applications, such as bulletproof vests and cut-proof gloves.

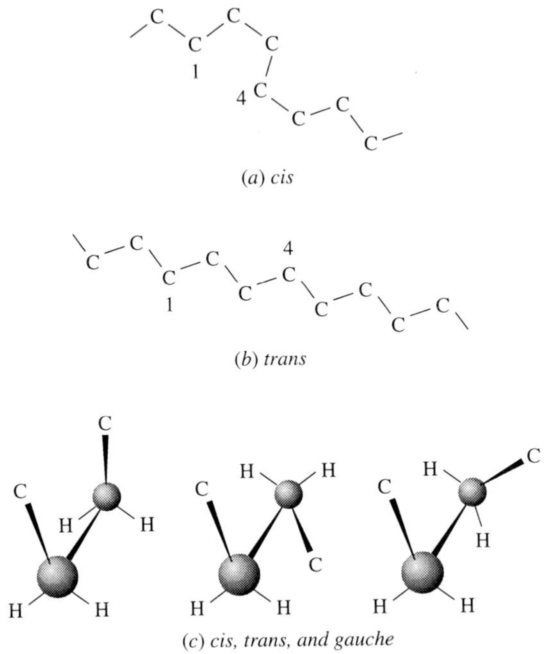

(a) cis

(b) trans

(c) cis, trans, and gauche

FIGURE 2.6 Conformations in polyethylene.

Because of the low mass of the mer and high molecular symmetry, ethylene must be polymerized to high molecular weights to have good physical properties. Textile grade PE is typically in the range of 200,000 g/mole, and industrial fiber may surpass 1,000,000 g/mole.

The final synthetic polymers used in textile applications that we'll discuss are elastomers. The most common elastomer is rubber, which may be either synthetic or natural. The term *elastomer* refers to the fact that the material can be stretched out many times its original length and recover almost completely when the force is removed. Vulcanized rubber is not a linear polymer, as are all the other polymers we have discussed so far. Rubber latex is a linear polymer, but to get the rubber molecules to return to their positions from a stretched state, the molecules must be imparted a memory. Thus, the molecules are crosslinked, usually in a process called vulcanization, as shown in Figure 2.7. Without the sulfur crosslinks introduced in vulcanization, rubber will not recover completely from large deformations. Crosslinks also ensure that a polymer is thermoset rather than thermoplastic.

A relatively recent invention in the area of elastomers is thermoplastic elastomers. Thermoplastic refers to the fact that these materials can be processed (repeatedly) by heating. A polymer that is not a thermoplastic is a thermoset. Thermosets cannot be processed repeatedly by melting. Vulcanized rubber is a thermoset polymer. Once vulcanized, rubber cannot be made to flow with the application of heat and pressure. (This is one reason automobile tires are

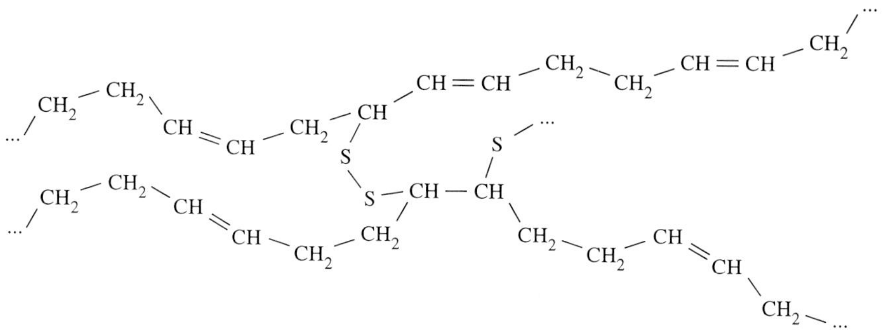

FIGURE 2.7   Structure of vulcanized rubber.

difficult to recycle.) Natural fibers such as cotton and wool also behave as thermosets.

Thermoplastic elastomers can have any of a variety of chemistries—polyesters, polyurethanes, polyolefins, etc. However, they are all block copolymers. A block copolymer is a polymer made up of two different sets of mers. The polymer is synthesized so that the similar mers react with one another. Two types of block copolymers are available, random block:

Aaaaaaabbbbbbaaaaaaaaaaaaaaabbbbbbbbaaa . . . A

and triblock:

Aaaaaaaaabbbbbbbbbbbbbbbbbbbbbbbbbaaaaaaaaa A

where the *a*'s and *b*'s represent different mers in the polymer chain. Both types of block copolymers function as thermoplastic elastomers. One block of mers is called a hard segment; it is rigid at room temperature. The other block of mer is called the soft segment; it is fluidlike at room temperature. The judicious combination of mers and the development of appropriate microstructure give these polymers their unique physical properties.

2.2   Chemical Structure of Natural Fibers

Nature has created fibers that are considerably more complex than the simple ones that man has synthesized. This is not surprising, in light of the fact that the fibers serve a variety of functions in the animal or plant. Plant fibers, such as cotton, linen, flax, and hemp, as well as wood fibers, contain mostly cellulose as the basic building unit, or mer. Animal fibers, on the other hand, such as wool, vicuna, or mohair, are based on amino acids. They are polypeptides, and the generic repeat unit is an amide. Let us first deal with cellulose.

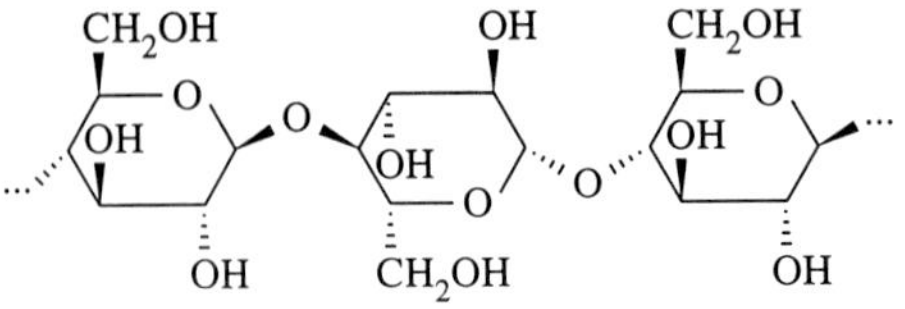

FIGURE 2.8 Chemical structure of cellulose.

2.2.1 Vegetable or Cellulosic Fibers

The structure of cellulose is shown in Figure 2.8. Note that the molecule is ribbonlike. Since it is a series of glucose rings joined together, cellulose resembles other vegetable-based polymers, such as starch. In starch all the hydroxyl groups are on the same side of the molecule, making it water soluble. The slight difference in bonding between starch and cellulose has profound effects on physical and chemical properties.

Cotton is essentially 100 percent cellulose, whereas wood is on the order of 50 percent cellulose. When wood is chemically pulped, the hemicellulose and lignin are removed, leaving only the tracheid fibers (cells), as shown in Figure 2.9, which are essentially entirely cellulose. What differentiates the various vegetable fibers is not so much the chemistry, but the microstructure and macrostructure of the fibers.

Cellulose has a molecular weight of about 10,000 g/mole. Because of its chemistry and molecular weight, cellulose is difficult to dissolve without degrading (cleaving bonds and reducing molecular weight). One technique for converting polymer into continuous fiber is to make a concentrated solution of polymer in solvent. To facilitate dissolution, cellulose is chemically modified. The modified cellulose can be dissolved, spun into fiber, the solvent removed, and the material optionally converted back to cellulose. Cellulose acetate, nitrate, and triacetate are products that may be produced from cellulose. In cellulose triacetate, for example, all the OH groups have been converted to

$$\begin{array}{c} O \\ \parallel \\ -C-O-CH_3 \end{array}$$

groups. The basic viscose process for making rayon, as shown in Figure 2.10, includes a step for converting the fiber back to cellulose, called regeneration.

The viscose process is not ideal: The regenerated cellulose has a molecular weight of less than 1,000 g/mole, and a great deal of environmental pollution (carbon disulfide and hydrogen sulfide) is produced in the viscose process. Hence, the cost of rayon has increased so much that the fiber has been largely displaced by other fibers. Research continues to be focused toward identifying good solvents for unmodified cellulose.

2.2.2 Animal or Polypeptide Fibers

As mentioned previously, all animal fibers have the same basic chemical structural units, polyamides, in spite of the fact that the fibers come from a variety

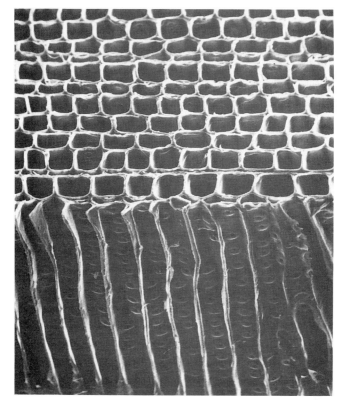

FIGURE 2.9   Structure of wood (jack pine) showing tracheids. (Courtesy of the Institute of Paper Science and Technology.)

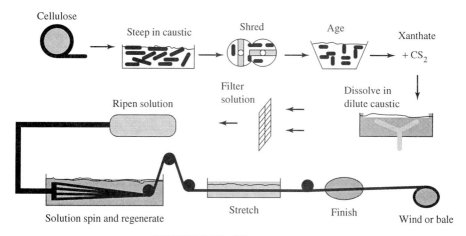

Cellulose

Steep in caustic

Shred

Age

Xanthate

$+ CS_2$

Ripen solution

Filter
solution

Dissolve in
dilute caustic

Solution spin and regenerate

Stretch

Finish

Wind or bale

FIGURE 2.10   Viscose process.

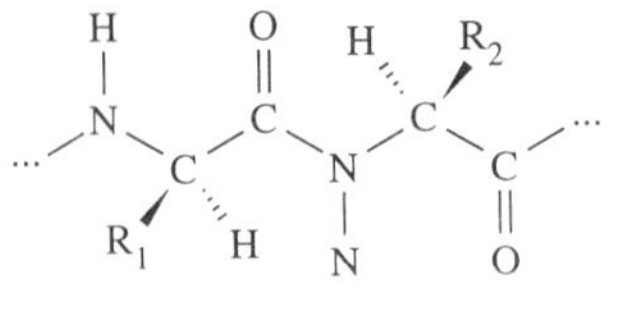

FIGURE 2.11 Structure of natural proteins.

of different animals. Wool is produced by sheep and silk by the silkworm or spider. The basic chemical repeat unit of animal fibers is shown in Figure 2.11 where the R_1 and R_2 groups are the amino acid residues. Table 2.1 shows the chemical characteristics of the two most important animal fibers to textile scientists, silk and wool. The chemical structure of silk is the relatively simple protein fibroin, which has a molecular weight of about 84,000 g/mole. That of wool is the protein keratin, which has a molecular weight of about 60,000 g/mole. Neither has ever been synthesized by man. Silk consists mainly of four

TABLE 2.1 Chemical Structure of Wool and Silk

Type	Side Group	Amino Acid	Silk Fibroin	Wool Keratin
Inert	—H	Glycine	43.8	6.5
	—CH$_3$	Alanine	26.4	4.1
	—CH(CH$_3$)$_2$	Valine	3.2	5.5
	—CH$_2$CH(CH$_3$)$_2$	Leucine	0.8	9.7
	—CH(CH$_3$)CH$_2$CH$_3$	Isoleucine	1.37	—
	—CH$_2$C$_6$H$_5$	Phenylalanine	1.5	1.6
Acidic	—CH$_2$COOH	Aspartic acid	3.0	7.27
	—CH$_2$CH$_3$COOH	Glutamic acid	2.03	16.0
Basic	—(CH$_2$)$_4$NH$_2$	Lysine	0.88	2.5
	—(CH$_2$)$_3$NHC(NH)NH$_2$	Arginine	1.05	8.6
	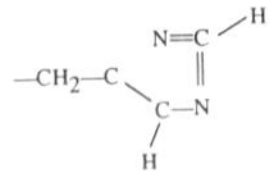 [imidazole structure]	Histidine	0.47	0.7
Hydroxyl	—CH$_2$OH	Serine	12.6	9.5
	—CH(OH)CH$_3$	Threonine	1.5	6.6
	—CH$_2$C$_6$H$_4$OH	Tyrosine	10.6	6.1
Ring	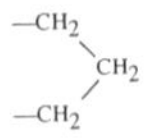 [ring structure]	Proline	1.5	7.2
Double	—CH$_2$—S—S—CH$_2$—	Cysteine	—	11.8
Other	—CH$_2$CH$_2$—S—CH$_3$	Methionine	—	0.35
	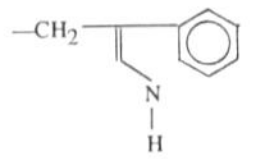 [indole structure]	Tryptophane	—	0.7

Source: M. Harris, editor, *Handbook of Textile Fibers*, Washington, D.C.: Harris Res. Labs, 1954.

short side groups. Wool contains many side groups, some large, and quite a number that can participate in intermolecular (between molecules) bonding. We will later focus on the importance of bonding between molecules in polymers. Note at this time:

1. The cysteine bond is a sulfur bridge between molecules, much like in vulcanized rubber. Wool contains about 12 percent cysteine; Angora rabbit, 14 to 15 percent; and human hair, 18 percent.
2. The acidic and basic groups can provide chain interaction by ionic coupling of salts.
3. The hydroxyl groups are capable of hydrogen bonding to the amide group.

What differentiates vegetable fibers from one another is chiefly the microstructure and macrostructure of the fibers. The same is true for animal fibers. We will address the similarities and differences in physical structure after discussing the nature of molecular bonding in textile fibers.

2.3 Bonding in Polymer Fibers

Bonds are the forces that hold atoms together and give materials integrity. We will discuss two classes of bonds, primary and secondary. Primary bonds are about an order of magnitude stronger than secondary bonds. Here we focus on primary bonds that bind polymers, the covalent bonds. They are about as strong as other primary bonds—metallic and ionic bonds. Only organic materials have strong secondary bonds. Without secondary bonding, polymers would be far less useful. Other than physical entanglements, interactions among chains would be similar to those between inert gas molecules.

2.3.1 Primary Bonding

It is clear from the discussions and figures that most textile fibers of interest to us are polymers that have strong primary bonds along the chain axis. The primary bonds in polymers are covalent bonds. In covalent bonds the atoms share electrons so that each atom has an essentially filled outer shell. The number of bonds to any atom is determined by the valence of the atom. Carbon, for example, will form four bonds, hydrogen will form one, nitrogen three, oxygen two, and so on. This characteristic may be compared to the other two types of primary bonds, metallic and ionic. In the case of metals, bonding is determined by how many atoms can be packed around each atom, often as high as 12. In ionic solids, bonding is determined by both the relative size of and the charge on each ion. Because the number of bonds is determined by valence in covalent structures, organics, which include polymers, are generally not very dense. In addition, polymers are composed of some of the lightest atoms in the Periodic Table. It is also worth noting that bonding in covalent structures occurs at certain angles that can be estimated on the basis of geometry alone. For example, carbon forms four bonds. Since bonds are regions

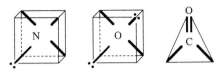

FIGURE 2.12 Geometry of covalent bonding.

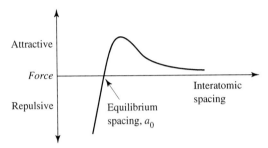

FIGURE 2.13 The effect of atomic separation on bonding force.

of high electron density, the bonds maximize their separation in space. Hence, the four bonds form a tetrahedron. Similarly, nitrogen forms three bonds, but a lone pair of electrons is present on the nitrogen, another region of high electron density. Consequently, bonding in nitrogen is also tetrahedral. Atoms with three bonds and no lone pairs will form bonds that describe an equilateral triangle, and atoms that form two bonds with no lone pairs form linear structures, as shown in Figure 2.12.

The strength of primary bonds—metallic, ionic, or covalent—is on the order of 300 to 500 kJ/mole. (This suggests that no one class of materials is inherently stronger than any other class.) A schematic of the force between two bonding atoms in a covalent structure is shown in Figure 2.13. The equilibrium spacing of atoms is where the attractive coulombic force between atoms is equal to the repulsive force, which is caused by electron overlap. Hence, covalent bonds have both characteristic bond angles and interatomic spacings. Table 2.2 lists various covalent bond specifications. They are also listed on the Periodic Table. Bond lengths of secondary bonds are longer, on the order of 3–5 Å.

2.3.2 Secondary Bonding

Organic materials are unique in that secondary bonding may be extremely important in spite of the fact that secondary bonds are about an order of magnitude weaker than primary bonds. Secondary bonding creates intermolecular interactions in thermoplastic polymers. Secondary bonds are the forces that bind collections of molecules, determine the amount of heat absorbed upon melting, and influence the crystalline melting temperature. They also determine in large part the interaction of the polymer with gases and liquids.

TABLE 2.2 Covalent Bond Specifications

Bond	Strength (kJ/mole)	Angle (deg)	Distance (Å)
C—C—C	368	109.5	1.54
C—H	435		1.09
C—O—C	360	108	1.43
C=O	531	120	1.23
C—N—C	305	109.5	1.48
N—H	460		1.01
O—H	498		0.97
C—N<	176		1.47
C—F	431	109.5	1.34
Si—O—Si*	368	142	1.50

Source: R. C. Weast, ed., *Handbook of Chemistry and Physics*, 63rd ed., Boca Raton: CRC, 1982.

*Not completely covalent: see text, pp. 25–26.

Types of Secondary Bonds

The strongest type of secondary bond is the hydrogen bond. Like all secondary bonds, the hydrogen bond is a result of charge interactions. Hydrogen bonding occurs between OH groups and between $>$C=O and HN groups. To understand hydrogen bonding, electronegativity must be understood. Some atoms have a greater affinity for electrons than do other atoms. Consider, for example, what occurs when Na and Cl atoms are brought together. Since Cl lacks one electron for a filled outer shell and Na has one more electron than a filled outer shell, the Cl has a high electron affinity and Na has a low electron affinity. When the two atoms approach, at some separation an electron from the Na jumps onto the Cl. Hence, the solid NaCl is really Na^+Cl^-, an ionic solid. Getting back to the hydrogen bond, we note that O is more electronegative than H, and N is more electronegative than H, due partly to the presence of lone electron pairs. This, together with the higher electron affinities of N and O than H creates more negative charge on the N and O than on the H. The N and O become partially negatively charged and the H becomes partly positively charged. Hence, the N and O are attracted to H's that are bound to other N or O. Hydrogen bonding is shown in Figure 2.14. In addition to organics containing OH, nylons, or polyamides, are capable of hydrogen bond-

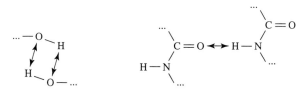

FIGURE 2.14 Schematic of hydrogen bonding in polymers.

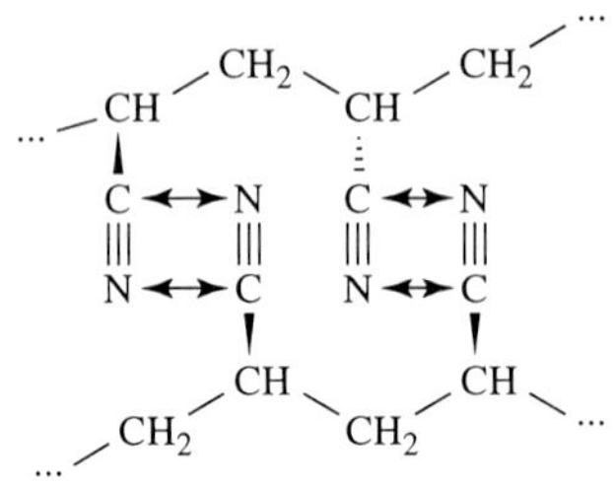

FIGURE 2.15 Secondary bonding in PAN by dipoles.

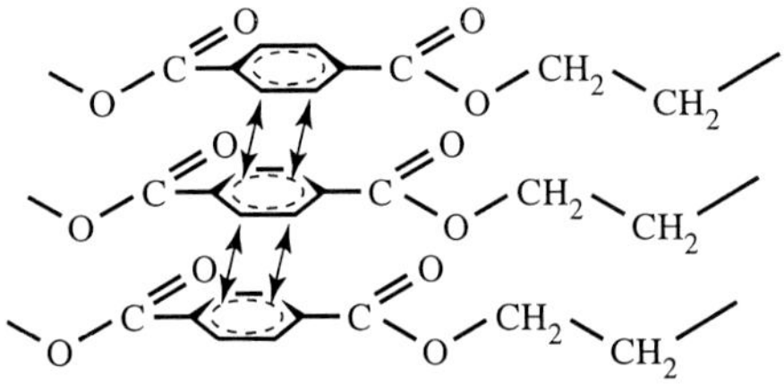

FIGURE 2.16 Secondary bonding by ring association induced by time varying charge distribution.

ing; polyesters are not. Materials that are capable of hydrogen bonding, such as natural fibers and synthetic polyamides, interact with atmospheric humidity and water.

Hydrogen bonds have strengths on the order of 40 kJ/mole. Other secondary bonds based on induced or permanent dipoles have strengths about half this value. An example of this type of bond is that in PAN, shown in Figure 2.15.

A third type of rather strong secondary bond is associated with aromatic structures. Aromatic rings are concentrations of somewhat mobile electrons. Because these charges can shift, rings are capable of electronic interaction and dipolar bonding. Rings in polymer crystals tend to stack regularly on each other, as shown for PET in Figure 2.16.

The weakest and final type of secondary bond is the van der Waals bond— 4 to 8 kJ/mole. Again the attraction is due to charge interactions between molecules; however, the charges cannot be readily identified. Van der Waals forces result from spacial charge fluctuations associated with the motion of electrons. As shown in Figure 2.17, polyethylene chains are held one to another by only van der Waals bonds. Hence, the melting temperature of PE is low.

Natural Fibers

Let us now examine the structure of cellulose, silk, and wool. Cellulose is capable of forming a number of hydrogen bonds. Silk and wool, too, can form hydrogen bonds. In fact, the hydrogen bonding in natural fibers is so extensive

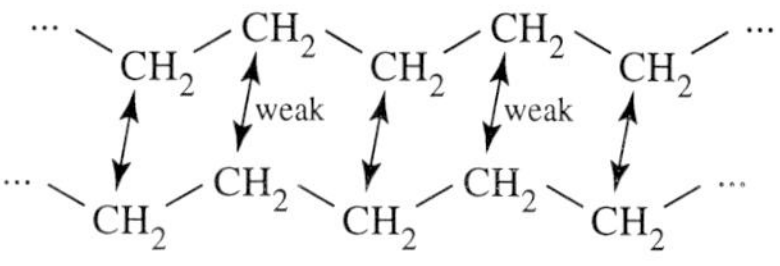

FIGURE 2.17 Weak van der Waals secondary bonding.

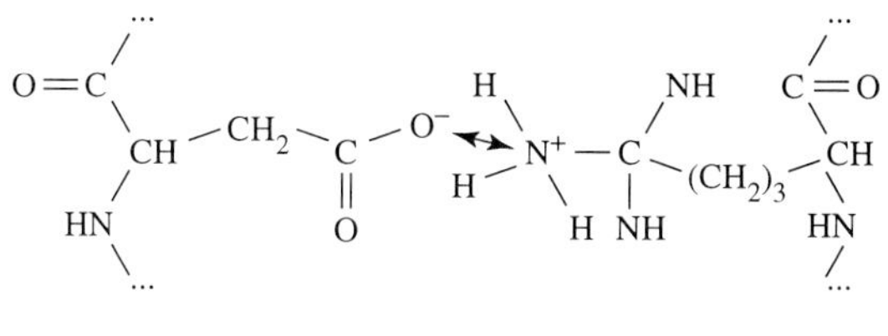

Salt of aspartic acid   Salt of arginine

FIGURE 2.18 Ionic bonding in natural polymers.

that these materials do not melt when heated; rather, the primary bonds break before all the hydrogen bonds can be eliminated. Hence, cotton, silk, and wool degrade before they melt. They behave like thermoset polymers, although they are not all crosslinked.

Wool can form not only hydrogen bonds, but also ionic bonds through the acidic and basic groups. These bonds are due to the attraction of permanently and oppositely charged moieties on the side groups, as shown in Figure 2.18. The ionic secondary bond is similar to the strong primary ionic bond, except that complete electron transfer does not occur. These ionic bonds may be extremely important. Consider, for example, a polymer fluid that contains both positive and negative ionic moieties, such as the natural polymer guar gum in water. Because of the association between the ionic groups, guar shows unusual rheological, or flow, properties. Guar is stiff and difficult to get to flow, but once it is moving, motion is easy to sustain. Hence, guar gum finds applications in enhanced oil recovery, where it is pumped into the ground and allowed to come to rest; then a shock wave is applied. The shock wave is transmitted by the guar to the rock, causing fracture of the rock. Guar gum is also a common additive to modern day ice cream. Such ice cream does not truly melt.

Wool also contains cysteine bonds, which are covalent or primary bonds bridging chains. Secondary bonds are temporary. They can be broken and reformed many times, but covalent bonds are permanent. Once a main chain bond is broken, the molecular weight is halved, and the bond will not reform under ordinary conditions. Melting or dissolution of a polymer, then, does *not* cause a reduction in molecular weight per se. Only temporary secondary bonds are disrupted. They can be easily reformed. A polymer with weak secondary bonds will melt or soften at a low temperature.

An example of the concerted strength of secondary bonding in cellulose is afforded by paper. An ordinary sheet of paper consists of cellulose fibers each only a few mm long. The fibers are held together solely by the concerted strength of a multitude of hydrogen bonds. No adhesive is used, yet paper shows remarkable strength and stiffness.

Neither primary nor secondary bonding need be pure. That is to say, bonding in a solid may be mixed, partly of one character, partly of another. An instructive example is silica, SiO_2. Fiberglass is a well-known, useful reinforcement material that is composed of about 70 wt percent silica. (E-glass, which is common, is a specific composition glass fiber.) The bonding in silica is between Si and O. Examination of the Periodic Table shows that O is more electronegative than Si. The difference in electronegativity, 1.54, however, is not sufficiently high that an electron from Si will completely transfer to the O. The chart of electronegativity difference on the Periodic Table shows indeed that silica is about 44 percent ionic and, hence, 56 percent covalent.

2.4 Summary

We have learned the basic chemical structure of a number of important natural and synthetic polymers that are used as fibers: cellulose, polyesters, polyamides, polyolefins, acrylics, polypeptides, vinyls, and the like. You should appreciate the basic chemical similarities and differences in the various animal fibers, especially silk and wool. Chain configuration, or tacticity, was presented. Syndiotactic, isotactic, and atactic configurations were shown.

Bonding between atoms and molecules was a key concept presented. Primary bonds are strong. Primary covalent bonds are highly unlikely to re-form once cleaved. Melting and dissolution do not affect primary bonds, only secondary bonds. Primary bonds are degraded at the temperature at which a polymer begins to change its chemistry. Secondary bonds are about an order of magnitude weaker than primary bonds, but they are nonetheless important since secondary forces are additive. Several types of secondary bonds were exemplified, from the strongest, hydrogen bonds, to the weakest, van der Waals bonds. You should be able to sketch a polymer chain and show both intramolecular (within a chain) and intermolecular (between chains) bonding. Secondary bonds influence many of the properties of polymers, such as glass transition temperature, melting temperature, melt viscosity, and solubility parameter, all of which we discuss in more detail in the chapters that follow.

References

W. von Bergen, ed. *Wool Handbook*, V1, 3rd ed. New York: Interscience, 1963.

S. Kumar. *Indian Journal of Fiber and Textile Research* 16 (1991), 52–64.

W. E. Morton and J. W. S. Hearle. *Physical Properties of Textile Fibers*, 3rd ed. Manchester, England: The Textile Institute, 1993.

L. Pauling. *The Nature of the Chemical Bond*, 3rd ed. Ithaca, NY: Cornell University, 1960.

F. Rodriguez. *Principles of Polymer Systems*, 3rd ed. New York: Hemisphere, 1989.

W. J. Roff. *Fibers, Plastics, and Rubbers.* New York: Academic Press, 1956.
S. L. Rosen. *Fundamental Principles of Polymeric Materials.* New York: Wiley-Interscience, 1982.
J. P. Schaffer, A. Saxena, S. D. Antolovich, T. H. Sanders, and S. B. Warner. *Materials Engineering.* New York: Times-Mirror Books, 1995.

Problems

(1) Use your knowledge of the chemistry of the interaction of water with polymers to predict whether water will interact with silica, SiO_2.

(2) Show with a sketch and discuss secondary bonding in: **(a)** PET, **(b)** mohair, **(c)** PAN, **(d)** i-PP.

(3) In product literature for a new fiber, you read that the material is a polyesteramide. What does this mean?

(4) What is the chemical structure of cotton? of wood pulp? of linen?

(5) Give two reasons why polymers are typically far less dense than most other solids.

(6) What are some of the important chemical differences between silk and wool?

(7) If the maximum temperature to which a polymer can be exposed is related to bond strength, what type of fibers do you anticipate will be most thermally stable?

(8) Discuss the importance of the slope of the force-separation curve (Figure 2.13) at the equilibrium position.

(9) A new animal fiber is brought to your lab for investigation. What do you anticipate regarding the chemical composition of the material? What if the fiber were a vegetable fiber?

(10) Many fibers are dyed using substances that form secondary bonds with the polymer molecules. Describe whether you expect acid dyes ($-SO_3H$ or their salts) to be effective at coloring cotton, wool, silk, and atactic polypropylene. Consider only affinity for the dye and rank the fibers.

(11) The melting temperature of a polymer is in part determined by the total strength of secondary bonding. Use your knowledge of chemical structure to predict which polymer in each pair has a higher melting temperature:
   (a) PP or PAN
   (b) PET or PBT
   (c) PVdF or PTFE (The former has 2 F, the latter 4 F, substituted for H in PE.)
   (d) nylon 6 or nylon 12

In Chapter 13 we will learn about other factors that affect the melting temperature of a polymer.

(12) Explain why wet paper has little strength.

(13) (a) Why are clothesline-dried cotton towels stiff and boardy? **(b)** Why do cotton fabrics wrinkle?

(14) For each of the following polymers, show and label all the forms of both intermolecular and intramolecular bonding: **(a)** PE, **(b)** rayon, **(c)** Vectra®, **(d)** PPTA, or Kevlar®.

(15) In the next chapter we will try to pack polymer molecules together as tightly as possible. How do you anticipate a-PP will pack? Compare with the packing of i-PP.

(16) Discuss the nature of bonding in a carbon fiber. Carbon fibers have a chemical structure similar to that of graphite, which consists of carbon as six membered rings, all connected or fused to give a planar structure. Within the plane the bond distance is 1.4 Å. The planes are stacked upon one another. In carbon fibers, the rings, planes, and stacking are imperfect. The fiber axis lies in the plane of the page, so ring stacking is normal to the fiber axis. The separation of the planes is about 3.4 Å.

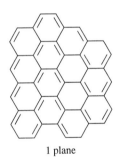

1 plane

(17) Describe the bonding in a ceramic fiber, such as Al_2O_3.

(18) PET, with a molecular weight of 20,000 g/mole, has how many mers?

(19) Why is the molecular weight of industrial fibers greater than that of textile fibers?

(20) Sketch the structure of Nomex®.

(21) Sketch the bond geometry associated with S and Si.

(22) Estimate the bond length, strength, and geometry of —C≡N.

(23) What type of primary bond does B form? (Consider, for example, B_2O_3.)

(24) On the basis of the chemical structure of wool and cotton, predict which fiber will have better recovery from stretching.

(25) Why do the molecules of wool form a helical path, like those described by Pauling for proteinaceous material, whereas those of silk are straighter?

(26) On the basis of chemical structure alone, will wool or cotton absorb more water?

3

Microstructure and Macrostructure of Fibers

Now that we understand fundamental aspects associated with the chemical structure of fibers, it is appropriate to address aspects associated with their microstructure. Microstructure is at least as important in determining properties as is chemical structure. Consider, for example, a fiber of polyethylene. Ordinary PE fibers have a strength of a few g/d; however, when the molecules are highly oriented and defects have been minimized, PE fibers may have strengths greater than ten times that of ordinary PE—more than 30 g/d.

In discussing the chemical structure of materials, we learned that primary and secondary bonds act over distances on the order of angstroms, Å (10^{-8} cm), and that polymer molecules may be as long as 100 μm (1 μm = 10^{-4} cm = 10^4 Å). When we begin to discuss the various aspects of physical structure, we again consider structures on the scale of atomic distances—on the order of angstroms, Å, or nanometers, nm (10 Å)—and we go up to distances on the order of cm, which is the length of most natural fibers. Aggregates of molecules or polymer crystals are generally in the range of 10's of Å to μm; and fibrils, which may be regarded as a fundamental unit of fiber structure, are typically 10 to 500 Å in diameter. Structure is important to physical properties on all scales, and we will address structural characteristics in the broadest sense of the word. Consider, for example, wool fibers. How many of you avoid wearing wool next to your skin because you find wool to be too scratchy? We will learn that the scratchiness of wool is a result of the molecular structure, conformation, and aggregation; the large diameter of the fiber; and the scales on the surface.

This is perhaps the most difficult chapter in the text. Many new terms and concepts are introduced. It is important that you grasp every one of them and understand the relationships among the concepts developed in the chapter.

3.1 Crystalline and Noncrystalline Materials

Let us begin our investigation of structure by introducing one of the most important concepts in the microstructure of solids, that of crystallinity. In order to understand the molecular organization of textile fibers, it is essential to understand the concept of a crystal. All organic textile fibers are semicrystalline. Textile fibers need to be tough, and the way to achieve toughness is to have crystalline regions. That is, textile fibers contain regions of both crystalline and noncrystalline material. Noncrystalline, or amorphous, materials may be either brittle solids, called glasses, or rubbery materials, called rubbers. Oxide glass (silica) is an inorganic material that is completely noncrystalline. Like other glasses, such as polymeric glasses, oxide glass is brittle. Rubbers, on the other hand, are easy to deform and fail at high elongation. Neither material is tough, meaning that only a small amount of energy or work is required to break the material.

One crude way that may be used to determine whether a polymer is semicrystalline or amorphous is to examine it visually. Semicrystalline polymers are always opaque in thick sections. Unfilled amorphous polymers are usually transparent and, hence, they may be used in fiber optics. The reason this test is crude is in part that filled amorphous polymers are usually opaque.

3.1.1 Structure of Crystals

To achieve an understanding of the nature of a crystal, it is perhaps best to discuss the simplest of all materials, pure metals. In an amorphous, or noncrystalline, material, the atoms are arranged in a random way, as shown in Figure 3.1 for a two-dimensional solid. In a crystal, on the other hand, the atoms are arranged on a periodic lattice. Each atom is located a precise distance in a precise direction from the others, also as shown in Figure 3.1.

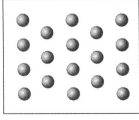
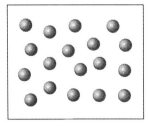

Crystalline   Noncrystalline

FIGURE 3.1   Structure of crystalline and amorphous materials.

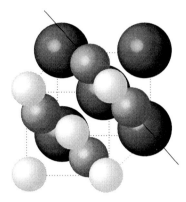

FIGURE 3.2 Unit cell of aluminum. The black atoms are on the back plane, the gray are on the midplane, and the white are on the front plane. The structure is face centered cubic. In real FCC solids atoms touch along the face diagonals. Note also the planes of close-packed atoms, such as the one projecting outward that includes the rear face diagonal shown.

An example of a 2-d crystal is wallpaper. Wallpaper has a motif that repeats periodically. To generate an infinite crystal of wallpaper, all that is required is the lattice and the repeat motif. The same is true of a 3-d crystal. Shown in Figure 3.2 is the unit cell of aluminum. The unit cell is the smallest representation of the entire crystal. Many properties of the crystal can be calculated by considering only the unit cell. Al has a face-centered cubic, or FCC, structure, which means that each cell contains 1/8 of an atom at each of the 8 cell corners and 1/2 an atom in each of the 6 face centers, giving a total of 4 atoms per cell. This is one of the simplest unit cells for any material, but it demonstrates all the salient aspects of a crystal:

1. A crystal has both short (1 to 3 Å) and long (1000 Å) range order.
2. Bond angles and atomic separations are maintained.
3. Only the lattice and the repeat unit are required to generate a crystal of infinite size.

Most materials encountered in ordinary life are almost completely crystalline. Important exceptions are oxide glass, which is window glass, and polymers, which are only partly crystalline at most. A definition of a glass is a noncrystalline material that is brittle. In sum, polymers may be glassy—brittle and noncrystalline—or they may be rubbery—flexible and noncrystalline. On the other hand, polymers may be semicrystalline, containing both crystalline and noncrystalline regions. Most semicrystalline polymers are tough.

3.1.2 Polymer Crystals

Polyethylene is chemically the simplest polymer, and PE crystals are the simplest polymer crystals. Polymer crystals are noticeably more complex than are crystals of pure metals. The unit cell of PE is shown in Figure 3.3. Clearly

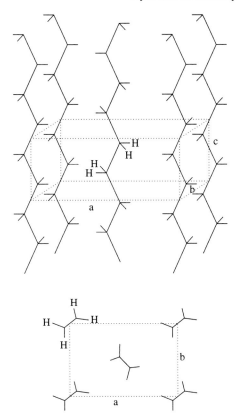

Top view

FIGURE 3.3 Unit cell of polyethylene; dimensions are a = 7.41 Å, b = 4.94 Å, and c = 2.55 Å. (C. W. Bunn and E. V. Garner, *Proc. Roy. Soc.* 47 (1947), A189.)

this cell is more complex than that of Al. Although all of the angles are right angles, the repeat spacing is different in the three principal directions, and several chains are present in a unit cell. Actually, the details of the crystal structure are not so important at this time as are some of the other features:

1. Crystals have both short- and long-range order.

2. Chains lie parallel and pack tightly in the crystal. Molecular orientation in most polymer crystals is high.

3. The unit cell is the smallest representation of the entire crystal. Consequently, many of the properties of the entire crystal can be calculated from the unit cell. For example, the density of the crystal may be calculated by counting the number of each type of atom in the unit cell, multiplying by the appropriate gram molecular weight, and summing over all atoms. The total weight is then divided by the volume of the unit cell to give the density.

The unit cells of nylon 66 and PET are shown in Figures 3.4 and 3.5. In the crystalline structure of nylon, note that hydrogen bonding is complete and no polar groups go "unsatisfied." In PET, the aromatic rings stack up to maximize electronic interactions (secondary bonding). The energy state of a crystal is always the lowest energy state of a solid material. Given sufficient time under appropriate conditions, materials that can crystallize do crystallize. The reason that polymers do not crystallize 100 percent and are instead semicrystalline is that the long chains are entangled in the melt and, upon cooling, the chains cannot disentangle sufficiently rapidly to crystallize. Growing crystals reject entanglements, making it impossible for the material between crystals formed in the initial stages of crystallization to crystallize. The reason that some polymers do not crystallize at all is largely that big, bulky, or asymmetric side groups cannot be incorporated into tightly packed, regular structures—crystals. An example already mentioned is atactic-PP, which is noncrystalline. Isotactic- and syndiotactic-PP are semicrystalline.

The crystal structure of PBT, poly(butylene terephthalate), is not at all like that of PET. The presence of the butyl sequence, four CH_2's rather than the two in PET, gives the crystal a helical conformation. (In general, large side groups require the molecular conformation to be helical to avoid steric overlap.) When a PBT crystal is stretched along the chain direction, the crystal structure changes, from what is called α to β. The length of the β cell in the chain direction is greater than that of the α cell, allowing the crystal to assume greater stretch. When the stress is removed, the crystal structure changes back to α, giving PBT very high recovery characteristics. Wool crystals behave similarly to PBT crystals under strain.

PAN is a polymer with a unique crystal structure. PAN is atactic, and the side groups are bulky and highly polar. Largely due to the bulky CN side groups, the molecular conformation is helical. PAN does not crystallize into a 3-d unit cell, but solid PAN has ordered regions that appear to have 2-d structure. The structure is similar to that of a bundle of pencils. The pencils' axis is the helix axis, and they lie parallel to one another and pack well together laterally; however, due to the fact that the polymer is atactic, there is no order along the length of the pencils—along the chain direction. The order in many liquid crystalline polymers is similar to that of PAN. Liquid crystalline polymers are generally stiff and rodlike, so the molecules lie parallel to one another, rather than coiled on one another as in a typical polymer melt or solution. The order of a liquid crystalline polymer in the fluid state is intermediate between that of a crystal and that of an isotropic, or disordered, fluid, hence the term *liquid crystalline.*

Crystallinity is not unique to synthetic polymers. Natural polymers may be semicrystalline. A good example is cellulose. When there are no defects in the chemical structure, cellulose may be 100 percent crystalline. Since natural polymers crystallize simultaneously with formation or synthesis, that is, as the chains grow, the entangled-in-the-melt arguments are inappropriate. Many

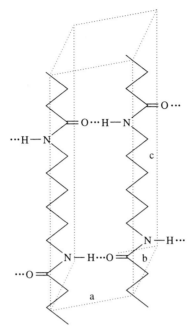

FIGURE 3.4 Unit cell of nylon 66. A molecule of nylon passes through each corner of the cell. Only two of the four chains are shown. Note that hydrogen bonding needs are satisfied. (C. W. Bunn and E. V. Garner, *Proc. Roy. Soc.* 47 (1947), A189.)

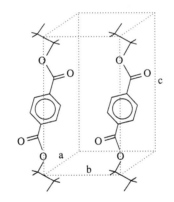

Top view

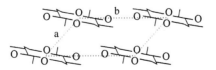

FIGURE 3.5 Unit cell of PET. A molecule of PET passes through each cell corner. Only two of the four chains are shown. Note that the aromatic rings are stacked. (R. P. Daubeny, C. W. Bunn, and C. J. Brown, *Proc. Roy. Soc.* 531 (1954), A226.)

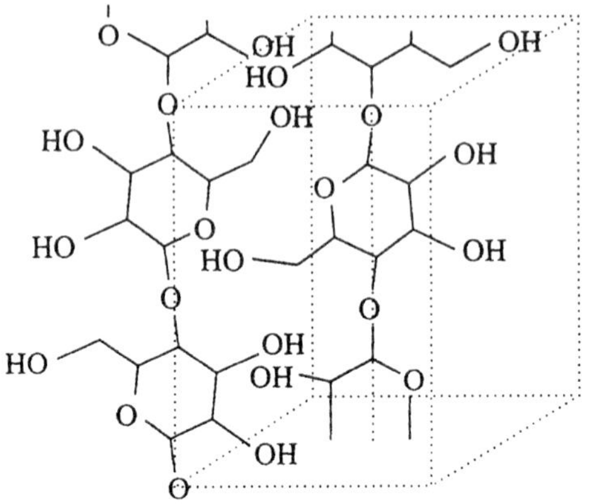

FIGURE 3.6 Structure of natural cellulose crystals. There are cellulose molecules passing through each corner of the cell and one through the center. Hydrogen bonding is maximized and the molecules pack parallel or antiparallel to one another. (K. H. Meyer and L. Misch, *Helv. Chim. Acta.* 20 (1937), 232.)

plant fibers are not 100 percent cellulose. Of the other two components that may be present in vegetable fibers, lignin is amorphous, and hemicellulose has low crystallinity. The structure of natural cellulose crystals is shown in Figure 3.6.

3.1.3 X-ray Diffraction and Crystallinity

By this time, you may have been wondering how scientists determine whether a material is crystalline or not, and if it is, whether it is completely or partially crystalline. A technique called X-ray diffraction is used to determine whether a sample is crystalline and to establish standards for assessing crystallinity.

To investigate the possibility of long-range three-dimensional order in a material, we need to probe the atomic positions in the material. To probe distances on the order of Å or nm with radiation, a characteristic wavelength on the order of Å or nm is required. Hence, X-rays are selected. The most commonly used source is Cu, and the wavelength of Cu radiation is 1.54 Å. A monochromatic collimated beam of X-rays is directed toward a sample, as shown in Figure 3.7. The X-rays interact with crystallographic planes in the material to produce a diffracted beam. (In the absence of crystals, there is no diffraction.) Diffraction occurs only under special conditions for each plane of atoms. For the planes shown in Figure 3.7, diffraction occurs when the path length difference between beam 2 and beam 1 is an integral multiple of the wavelength, i.e., when the beams that appear to be reflected beams are in-phase:

$$n\lambda = 2(d\sin\theta)$$

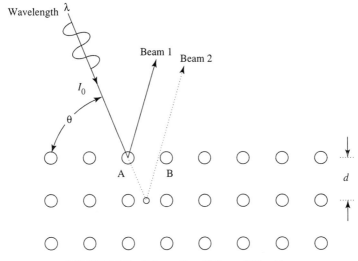

FIGURE 3.7 Schematic of X-ray diffraction.

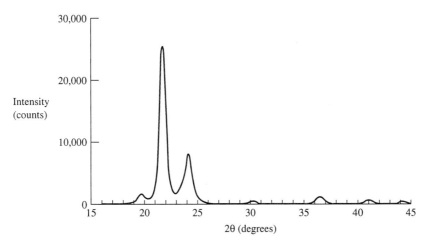

FIGURE 3.8 X-ray diffraction from polyethylene crystals.

where $d\sin\theta$ is the length of segment A or B in Figure 3.7. This equation is Bragg's law, and it describes the conditions for diffraction. At the appropriate angle, diffraction from a set of planes occurs. Since an unoriented sample contains many sets of planes at different spacings and orientations, a scan of intensity vs. angle (θ) gives a number of peaks, as shown in Figure 3.8 for PE. The intensity and position of the peaks can be calculated from knowledge of the crystal structure, or vice versa. In the absence of crystals, there are no peaks. Crystallinity, then, is determined by measuring the area under the peaks and comparing the values with those for the pure crystal, which are either calculated or measured.

X-ray diffraction is rich in techniques, and a wealth of information can be learned from each one. We are concerned at this time with only a few simple aspects of X-ray diffraction. X-ray diffraction can be used to determine:

- Crystal structure. From the crystal structure, we can calculate the density of the crystal.
- Percent crystallinity. X-ray diffraction provides the only absolute measure of polymer crystallinity.

We can use other techniques to determine the percent crystallinity in a sample only after we have established a set of "standard" samples covering a range of crystallinity.

3.1.4 Assessment of Crystallinity

Three techniques are commonly used to determine crystallinity:

1. X-ray diffraction.
2. Heat of fusion.
3. Density studies.

In X-ray diffraction a peak area or height is analyzed. The higher the peak or the more area under the peak, the more crystalline the sample. Even better, an average of a number of peak areas can be used.

The technique most often used to assess heat of fusion in a fiber is DSC, or differential scanning calorimetry. It is a rapid technique, employing a heating rate of about 20 °C/min, and modern instruments calculate the heat of fusion per unit mass of material, as shown in Figure 3.9. Specifically, a sample is melted in a calorimeter under controlled conditions in heat-of-fusion studies. At the crystalline melting temperature, excess heat is pumped into the sample to effect the transition from crystalline solid to melt. The amount of heat required is proportional to the fraction of crystal present, to the crystal being melted, and to the amount of sample being heated. The heat of fusion per unit mass of pure crystal must be known to assess crystallinity. These data are available for various materials and have been compiled in Table 3.1. To calculate crystallinity, χ, take a ratio of the heat of fusion per gram of sample to that of the pure crystal:

$$\chi = \Delta H_f \text{ (per g of sample)}/\Delta H_f \text{ (per g of crystal)}$$

The third technique for estimating crystallinity is based on density. Implicit in this technique is the approximation that the material is two-phase, consisting

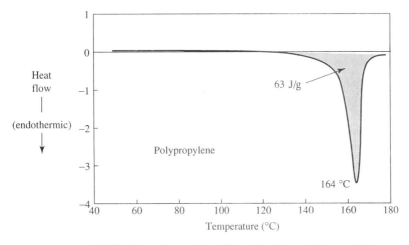

FIGURE 3.9 DSC of a polypropylene fiber showing melting endotherm.

of crystalline and amorphous regions. For most polymers, this approximation is reasonable, and we will use it throughout the text.

The crystal density may be calculated from X-ray diffraction data. The density of amorphous polymer may be measured on a sample with no crystallinity, calculated from density measurements on a semicrystalline sample, or extrapolated from values in the melt. Fiber densities range from that of amorphous to that of crystalline material:

$$\text{crystallinity, } \chi = (\rho_{\text{fiber}} - \rho_{\text{am}})/(\rho_{\text{cr}} - \rho_{\text{am}})$$

The densities of various fibers are given in Table 3.2. Also included in the table is the density of crystalline and noncrystalline polymer.

TABLE 3.1 Heats of Fusion of Polymer Crystals

Material	Heat of Fusion (J/g)
cellulose	does not melt
polypeptides	do not melt
nylon 6,66	230, 300
PET	140
i-PP	165
PE	293
PTFE	68.3
1,4 polybutadiene	170

Source: B. Wunderlich, *Macromolecular Physics: Crystal Melting*, V3, New York: Academic Press, 1980.

TABLE 3.2 Densities of Fibers and Crystals

Fiber	Typical Fiber	Crystal	Amorphous
		Densities (g/cm^3)	
Cotton	1.52	1.60	1.49
Wood pulp	1.54	1.60	1.49
Viscose rayon	1.49	1.60	1.49
Triacetate	1.32		
Wool	1.31		
Silk	1.34	1.37^b	1.29*
Nylon 6,66	1.14	1.23, 1.24^a	1.10
PET	1.39	1.455*	1.335
Acrylic	1.18	(2-dimensional order)	
PP	0.93	0.936^a	0.85
PE	0.96	1.00^a	0.86*
PTFE	2.20	2.30^a	2.00
PPTA (Kevlar® 149)	1.47	1.50^a	
PMIA (Nomex®)	1.45	1.45^a	
PBO	1.58	1.67^c	
Carbon (AS-4)	1.80	(2.25 graphite)	
Oxide glass (e-glass)	2.62	(2.65 quartz)	2.62

Values at 20 °C and 65% relative humidity

Sources: J. Brandrup and E. H. Immergut, eds., *Polymer Handbook*, 3rd ed., New York: Wiley-Interscience, 1989; W. E. Morton and J. W. S. Hearle, *Physical Properties of Textile Fibers*, 3rd ed., Manchester, England: The Textile Institute, 1993; ^{a}H. Todokoro, *Structure of Crystalline Polymers*, New York: Wiley-Interscience, 1979; ^{b}D. L. Kaplan in *Biomaterials: Novel Materials from Biological Sources*, New York: Stockton Press, 1991; ^{c}S. Kumar, "Advances in high performance fibers," *Indian Journal of Fiber and Textile Research* 16 (1991), 52–64.

*Estimated from reported values of crystallinity, fiber density, and crystal density.

The three techniques described above each have inherent weaknesses and limitations. We have ignored interfaces between crystalline and noncrystalline regions, which may be rather extensive; the effect of orientation on density, heat of fusion, and X-ray intensity; the possibility of more than two phases being present, such as some voiding, or two crystalline or amorphous phases; etc. In practice the three techniques rarely give numbers that agree in detail; however, the numbers are nonetheless useful, since many physical and chemical properties depend on whether the molecules are in crystalline or noncrystalline regions. Use the values only as estimates. The techniques are not capable of producing accurate values for crystallinity. (For this reason I have not mentioned the difference between mass- and volume-based crystallinity.)

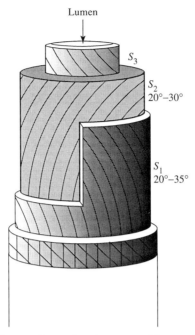

Lumen

S_3

S_2
$20°$–$30°$

S_1
$20°$–$35°$

FIGURE 3.10 Molecular conformation in cotton fiber and layering. (R. Jefferies, D. M. Jones, J. K. Roberts, K. Selby, S. C. Simmons, and J. O. Warwicker, *Cell. Chem. Tech.* 3 (1969), 255.)

3.2 Microstructure and Macrostructure of Natural Fibers

Man cannot control the structure of natural fibers. What you harvest is what you get. Man's control of fiber properties is on the species that he raises. This makes it somewhat easier to describe the generic morphology of natural fibers than that of synthetic fibers, yet it makes it difficult to specify the physical properties.

Macrostructure is the structure of a material as seen by the naked eye. Microstructure is the structure of a material as observed using a microscope. Both microstructure and macrostructure reflect the physical arrangement of groups of atoms. Neither is concerned with the chemistry of the material per se, although chemistry may influence the physical structure.

3.2.1 Cotton and Other Vegetable Fibers

Cotton is chiefly cellulose with various other substances at or near the surface, such as pectins, fats, and waxes. The cellulose molecules in cotton, as in other plants, spiral. The helix angle in cotton is 20 to 30 degrees. The molecules in cotton run parallel to one another, spiraling about the axis in layers, as shown schematically in Figure 3.10. When growing cotton is viewed on-end,

FIGURE 3.11 Swollen cotton fiber growth rings. (T. Kerr, *Protoplasma* 27 (1937), 229.)

however, the layers are not seen. Rather, growth rings are observed, much as in a tree, as shown in Figure 3.11. Cotton fibers are on average 33 mm long. Cotton is a hollow fiber when growing, but the lumen collapses when the fiber dies or is harvested, producing a flat fiber, as shown in Figure 3.12. The flat fiber twists, giving cotton fiber natural texture. When cotton is viewed on its side (transversely) in a transmitted-light microscope at high magnification, or when fracture ends are examined, cotton usually appears to be composed of microfibers called microfibrils, as shown in Figure 3.13.

Cellulose molecules can form highly crystalline polymer. Crystallinity in cotton is thought to range from 60 percent to as high as 100 percent. Cotton crystallinity is 60 percent for those who envision crystals to be nearly perfect and 100 percent for those who believe the crystals to be highly defective. In this course we will consider cotton to be essentially completely crystalline, the crystals being highly defective. Ramie, flax, and hemp contain a lower percent of cellulose than does cotton. In general, these fibers are multicellular, and the helix angle of the cellulose is lower than that of cotton—6°, compared with cotton's 20 to 30°. Among other effects, a lower helix angle increases the stiffness of the fiber and makes the fiber brittle. This is readily understood by comparing the mechanical properties of a solid steel wire with those of a coil spring.

3.2.2 Silk, Wool, and Other Animal Fibers

We have discussed the chemical structure of wool and silk as well as the nature of secondary bonding in polymers. Recall that both silk and wool can form many hydrogen bonds, but only wool can form salt linkages and only wool has the covalent cysteine crosslinks bridging molecules.

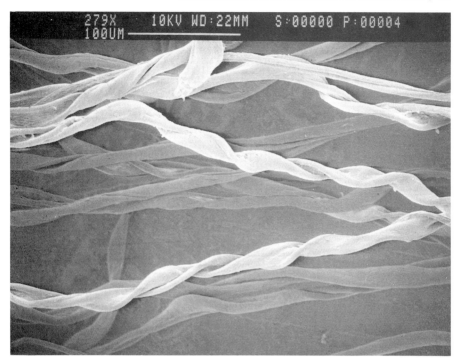

FIGURE 3.12   The texture of cotton fiber.

The molecules in silk, chiefly fibroin, are parallel to the fiber axis and they are stretched out to form what are called partially extended chain crystals, as shown in Figure 3.14. The sections of the silk molecule that have bulky side groups generally lie in the noncrystalline regions, since it is difficult to align the bulky side groups in periodic structures. The organization of crystalline and noncrystalline regions is similar to that in cellulose-based natural fibers; i.e., a micellar model may be used to depict the structure, as described in detail at the end of this chapter. In cross-section silk can be envisioned roughly (very roughly) as two triangles of fibroin glued together with the noncrystalline protein sericin. Silk is a straight fiber, lacking scales and texture along its length. Silk has for centuries been valued for its properties, and man continues to try to make a synthetic fiber that reproduces all the properties of silk (M. Fukuhara, *Textile Research Journal* 63 (1993), 387.)

The molecules in wool, which are chiefly keratin, form helices with a characteristic helix angle of 30 to 35°. The sections of helically oriented chains with small side groups lie within the crystalline regions. When a stretching force is applied to wool fiber, the crystal structure reversibly changes from α to β. The β-crystalline form has a longer unit cell in the chain direction than does the α-form. Thus, the transformation relieves some of the stress, allowing wool to endure higher elongation than might otherwise be possible. The

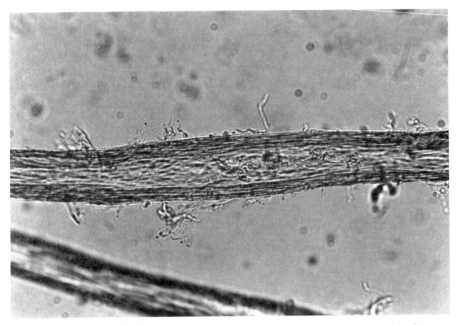

FIGURE 3.13 The fibrillar structure of cotton. Sample was prepared by shearing wet cotton fiber in a mortar and pestle. Examination is by transmitted polarized light of fiber in water.

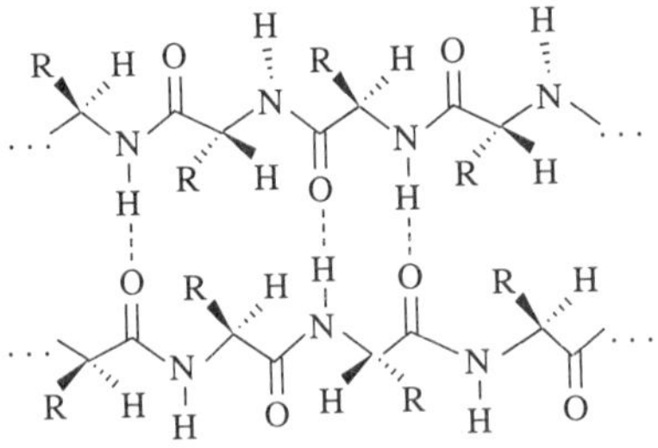

FIGURE 3.14 Conformation of molecules in silk crystals (fiber axis horizontal).

molecules in wool fold and coil, yet at high elongation the molecules tend to be stretched out; however, the cysteine crosslinks provide a memory of the original low-energy molecular positions and hence facilitate good recovery from deformation. Wool is roughly 40 percent crystalline. Wool fibers are almost round in cross-section and modeled with the structure shown in Figure 3.15. Note that wool consists of many cortical cells running parallel to the fiber axis and cuticular scales covering the surface of the fiber. The scales are important in the frictional properties of wool, allowing the fibers to slide past one another much more easily one direction than in the other. The scales allow

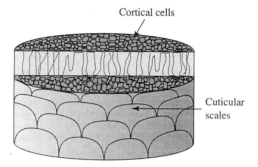

Cortical cells

Cuticular
scales

FIGURE 3.15   Structural model of wool fiber.

wool to felt, which is basically just irreversible fiber entanglement, usually of carded bats.

Wool fibers may be large (40 μm diameter as in English Leicester), medium (30 μm as in Corriedale), or small (20 μm as in Merino) depending on the breed of sheep. Sheep are shorn when the wool reaches a length of 5–20 cm. The optimal staple length varies inversely with the fiber diameter. Wool has texture not only from the scales, but also from the natural crimp, as shown in Figure 3.16. Although round in cross-section, wool fibers are not radially symmetric in cortical properties. There are two sides, an orthocortex and a paracortex, that have slightly different compositions, structures, and properties. The crimp is due to the effect of environment on the length of each side, bending much like a bimetallic strip bends with heat. Hence, crimp frequency varies inversely with fiber diameter.

The importance of cysteine bonds in wool cannot be understated, and we will direct your attention to the cysteine bond to rationalize properties countless times in this text. Wool and human hair are similar in structure and composition, so knowledge of wool translates directly into knowledge of human hair. When people have their hair "permed," what is being done? In order to permanently deform hair, the cysteine bond, a primary bond, must be broken. Hence, the chemicals used in the perm break the cysteine (sulfur) bond. The hair is then put into the desired position and some of the cysteine bonds are hopefully re-formed using the chemistry of the perm. Perhaps you can now understand why a permed head of hair smells a bit like vulcanized rubber!

3.3 Microstructure and Macrostructure of Synthetic Fibers

To understand the structure of man-made fibers, it is necessary to understand the microstructure of semicrystalline polymers. We will therefore briefly review crystallization in polymers from the melt or concentrated solution. Then we will discuss what occurs in fiber formation, which necessitates a presentation of fiber extrusion techniques. Finally we will get to the task at hand and

(a)

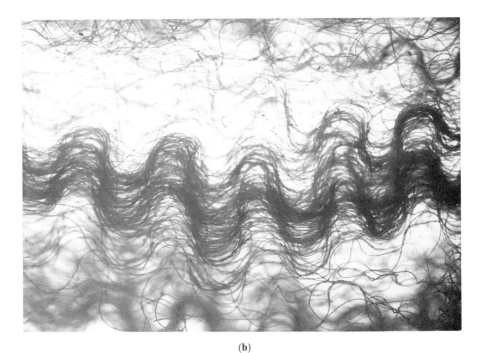

(b)

FIGURE 3.16   (a) Scales and (b) crimp in wool fiber.

discuss the structure of fibers, focusing on orientation phenomena. The chapter closes with a description of models used to describe the arrangement of crystalline and noncrystalline regions in fibers.

3.3.1 Polymer Crystallization from the Melt or Concentrated Solution

Consider a melt of polymer chains. Flory showed that the lowest energy state of the chains in a melt is when they are intertwined and coiled on themselves, much as spaghetti on a dinner plate. Upon cooling to below the melting temperature, the lowest energy state becomes the crystal. We have learned that the chains in a polymer crystal are parallel and not entangled. The process of reordering the chains from their entangled state in the melt to their stretched-out or helical state in the crystal takes much time. In practice a polymer must settle for a compromise. Under quiescent conditions (no stirring, no flow, no molecular orientation), the compromise is spherulites. Spherulites are aggregates of crystalline and noncrystalline regions. They grow radially in all directions from nuclei. The chains are oriented normal to the radial direction. The crystals in the spherulites, called lamella crystals, are small and the chains fold irregularly. A photomicrograph of spherulites is shown in Figure 3.17; a sketch of a spherulite's structure is also shown. Spherulite radius grows linearly with time in virtually all polymers crystallized under quiescent conditions, as shown in Figure 3.18a. The linear growth rate is a function of temperature, as shown in Figure 3.18b. The maximum in the curve is a result of high nucleation rate, which is formation of growth centers, at low temperature or high undercooling $(T_m - T)$, and high growth rate at high temperature or low undercooling. Again, essentially all polymers show this behavior. In fact, virtually all materials that crystallize are characterized by a maximum rate determined by competition between nucleation rate and growth rate.

Let us return to the time-dependent development of crystallinity. In the melt the crystallinity is zero. When the polymer is cooled and held isothermally at the appropriate temperature, nuclei form and grow as spherulites until the spherulites impinge, or grow into one another. The rate of growth is similar to that in grains of other materials, and the kinetics of growth are similar. Johnson and Mehl and Avrami quantified the rate of crystal formation (Sharples, 1966):

$$\chi = 1 - \exp(-kt^n)$$

where

χ = time-dependent volume fraction crystallinity,

k = Avrami rate constant, which is a constant at any temperature, and

n = Avrami exponent or shape parameter, which gives information on the type of nucleation and shape of crystal growth, such as disc-shaped, linear, or spherical.

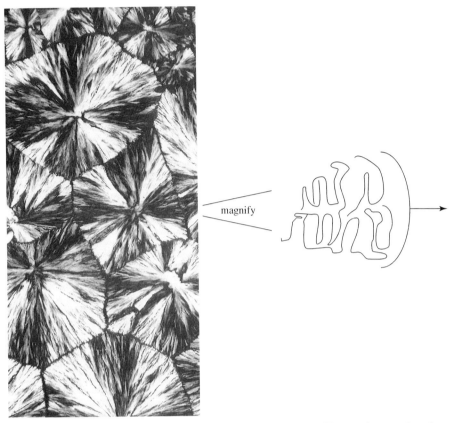

FIGURE 3.17 Appearance and structure of spherulites. (Photomicrograph taken under crossed polars.)

Data that facilitate calculation of Avrami parameters may be derived from studies conducted during crystallization using dilatometry, X-ray diffraction, density, or thermal analysis (DSC). For the past couple of decades, DSC has been by far the most widely used technique in crystallization studies of polymers.

Spherulites form under only quiescent conditions. What is the nature of crystallization from a melt in which the molecules are oriented, lying essentially parallel to one another, rather than entangled and coiled on themselves? Recall that the major barrier to crystallization is to arrange the chains as they appear in crystals. With aligned chains in the melt, the major barrier has been removed and crystallization occurs quickly and easily. In fact, estimates show that crystallization may proceed up to 10^{100} times faster! Since we cannot achieve perfect alignment and disentanglement of all molecules, non-crystalline regions are also present in solidified oriented polymer; however, the structure is not spherulitic. Rather, the crystals tend to be elongated (row). Often an overgrowth grows on the elongated crystals, giving what

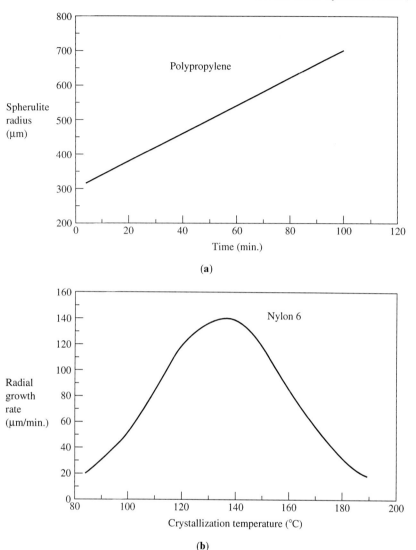

FIGURE 3.18 (a) Radial growth rate of polypropylene spherulite at 125 °C. (H. D. Keith and F. J. Padden, Jr., *J. Appl. Phys.* 35 (1964), 1270.) (b) Growth rate of spherulites of nylon 6. (J. H. Magill, *Polymer* 3 (1962), 655.)

is known as a row-nucleated structure, as shown in Figure 3.19. The row (bright line) is partially extended chain crystal, and the overgrowth is lamellar crystal. Spherulites and row-nucleated structures are extremes in polymer crystallization from the melt or concentrated solution. In practice most real processing falls somewhere between these two extremes. Let us consider typical fiber formation techniques.

FIGURE 3.19 Row-nucleated polypropylene.

3.3.2 Fiber Formation

Fibers may be formed using a number of techniques that may be loosely categorized as either solution or melt processing techniques. Melt processing is the easiest to understand, so we will consider that first.

In melt spinning, as it is called, polymer pellets are gravity fed into one end of an extruder. The screw engages the pellets and carries them into the barrel, where the pellets are subjected to heat and shear and consequently soften. The screw action delivers the mixed and molten polymer to the spinneret, which has up to 1000 shaped holes for fiber formation. A molten stream of polymer is forced by pressure through the shaped holes, whereupon the extrudate is stretched and solidified. Attenuation, or diameter reduction, and molecular orientation are achieved by taking up the fiber many times, perhaps 100 times faster that it is extruded. After the fibers are solidified by cooling with flowing air, they are subsequently drawn or stretched to further increase the molecular alignment in the axial direction. Note that crystallization occurs in an oriented melt, so no spherulites are formed using this technique. The melt spinning process, the most economical fiber formation process, is sketched in Figure 3.20.

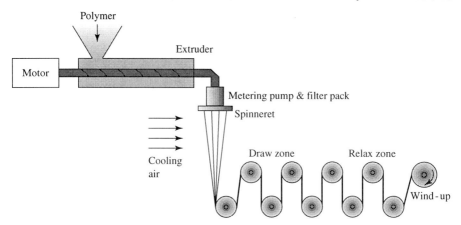

FIGURE 3.20 Schematic of melt spinning process.

In solution spinning techniques, shown in Figure 3.21, polymer is dissolved in high concentration in a solvent that does not degrade the polymer. The viscous dope is supplied to the spinneret under pressure and extrusion occurs. In dry spinning the extrudate is stretched and air is blown over it to cause evaporation of the volatile solvent. The solvent is recovered and the fiber may be further stretched or processed and finally wound on bobbins. In wet spinning extrusion occurs into a nonsolvent. The fiber is stretched while the solvent diffuses out and nonsolvent diffuses into the fiber, effecting precipitation. The washed fiber is further stretched, treated, and wound onto bobbins. Again, note that crystallization occurs in an oriented, concentrated solution, so no spherulites are formed.

The fiber skin is the first region to establish in solution spinning; however, the fiber at this time is swollen with a large amount of solvent. Consequently, the cross-sectional area encompassed by the skin is too small for the fiber to remain round after solvent removal. Hence, solution spun fibers are not round. If they appear round, then the surface is crenulated. Often the cross-section is bilobal or trilobal:

Crenulated   *Bilobal*   *Trilobal*

Most fiber spinning processes are one of the three just given; however, there are mixed processes specially designed for various materials. PPTA, for

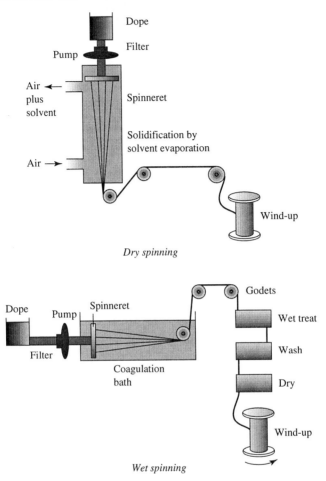

FIGURE 3.21 Schematic of wet and dry spinning process.

example, must be spun from hot solution, and the coagulation bath is most effective when cold. Hence, the spinneret is removed from the bath in a process called dry jet wet spinning. Acrylic polymers cannot be melt spun, so they are usually dry or wet spun. Acrylic polymer can be combined with water under pressure and the solution can be semi-melt spun at elevated temperature. In general, melt spinning is more economical than wet or dry spinning, chiefly because no solvent recovery is involved.

In man-made fibers the fiber cross-section is determined by the shape of the spinneret hole. Fiber may be round, oval, hollow, or virtually any shape. Shape is selected on the basis of desired optical, electrical, thermal, mechanical, or other properties. Man-made fibers do not have a characteristic shape as do natural fibers. Man-made fibers have no set diameter or cross-sectional area. If a large-diameter fiber is required, the spinneret hole is increased, the draw is reduced, or both. (Large-diameter fibers are useful, for example, in the carpet industry. Small-diameter fibers are used, for example, by the apparel industry in applications against skin because they are soft.)

Man-made fibers are formed on a continuous basis. Staple fibers are a result of deliberate cutting, in hope of mimicking the tactile and aesthetic properties of natural fibers. Man-made fibers generally have no texture, as do wool or cotton fibers. Texture is imparted in a separate step called texturing or bulking. The purpose of texturing is to change the fiber from being a straight rod to being a crimped rod so that synthetic fibers are given a contour roughly similar to that of wool. Texture can be helical or zigzag.

3.3.3 Microstructure of Man-Made Fibers

We have learned how polymer is converted into fiber. What is the structure of the fiber? Like natural fibers, man-made fibers are semicrystalline. They can be approximated as consisting of two phases, a crystalline phase and a noncrystalline phase. The noncrystalline, or amorphous, phase is an entangled, low-density mass of polymer chains. The crystals are relatively small, high-density units that may or may not be extended in the axial direction, depending on processing. Of key importance to fiber and textile scientists are the size, shape, and connectivity of the phases, and the orientation of the crystals and the molecules in the noncrystalline regions.

Let us begin by discussing the orientation of molecules in the noncrystalline regions and its importance. Molecular orientation is a theme that appears numerous times in this text, for example, when we discuss fiber shrinkage and thermal properties, birefringence, mechanical properties, and density. It is one of the most important properties of a fiber. Without molecular orientation, synthetic organic fibers are weak, they may stretch many times their original length in response to a small applied force, and they have poor recovery from deformation. They would be inadequate in most industrial and textile fiber applications. In addition, some polymers, such as PET and nylon, essentially do not crystallize without orientation. Thus, it is imperative that organic fibers be imparted orientation in processing.

Suppose a fiber is produced as described in the fiber-formation section of this chapter. Some orientation is imparted in spinning, but the lion's share of the orientation is imparted in separate drawing steps conducted on solid fiber. In the drawing steps the fiber may be drawn at or near room temperature, or at higher temperatures, or both. Molecules may be pulled out of the lamellar crystals and molecules in the noncrystalline regions are extended, as shown in Figure 3.22. During drawing, crystals align with the stretch direction, thereby aligning the molecules within the crystals with the stretch direction. The net result is high molecular orientation in both crystalline and noncrystalline regions.

The chief problem with this structure as a textile fiber lies with the noncrystalline regions. When these molecules are imparted sufficient heat that the

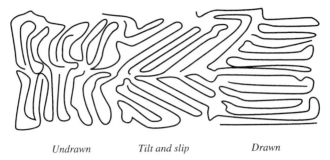

Undrawn   *Tilt and slip*   *Drawn*

FIGURE 3.22   Effect of drawing on molecular arrangement.

chains become mobile, they seek to coil on themselves, causing considerable fiber shrinkage. To avoid this problem, textile fibers are given a relax step prior to winding on the bobbin. The relax step consists of heating the fiber to above the temperature where molecules in the noncrystalline regions become mobile and allowing the fiber to shrink a few percent. Some additional crystallization may occur, and the molecular orientation in the noncrystalline regions decreases. Fibers treated in this fashion are stable to washing temperatures and to dry heat. In addition, decreasing the orientation of the molecules in the noncrystalline regions increases the diffusion rate of dyes and allows these regions to accept dye more uniformly than when the molecules in the amorphous regions are highly oriented.

3.3.4 Models of Fiber Structure

We are now ready to present models for the structure of fibers. Note that all the models for fiber structure presented at this time are two-phase models, the phases being crystalline and noncrystalline material. Most of the time, two-phase models are satisfactory, but not always.

The fringed micelle model is one of the oldest models and still enjoys considerable popularity. Shown in Figure 3.23 is a fringed micelle model for an unoriented polymer. Most, if not all, textile and industrial fibers are highly oriented, and they are characterized by a fibrillar microstructure. Consequently, the fringed fibril structural model was developed to model fiber structure; it is shown in Figure 3.24. Fibrils are clusters of partially aligned molecules. Fibril diameters range from 10's of Å to 1000 Å. Forces between fibrils are weak, so fibrils can often be observed in fracture studies on oriented fibers, as shown in Figure 3.25. Fibrils are an important aspect of fiber structure. They exist in both natural and synthetic fibers. Microfibrils, the most elementary fibrils, consist of alternating layers of ordered (crystalline) and disordered (noncrystalline) regions along the fibril length. The major criticism of this model is that some chain bending and folding, albeit irregular, occurs at the crystal surface. Hence, a modified fringed micelle (or fibril) model, as shown in Figure 3.26,

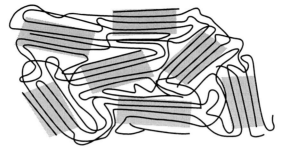

FIGURE 3.23   Fringed micelle model. (Crystalline regions are shaded.)

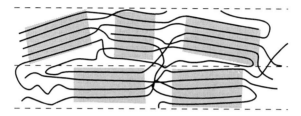

FIGURE 3.24   Fringed fibril model of fiber structure.

more accurately depicts the structure of semicrystalline polymers. This model is a reasonable model of unoriented semicrystalline polymers.

The final model that we will discuss is the Peterlin model, shown in Figure 3.27. This model was developed to show the structure of drawn polymers, so it seems to be tailor-made for fibers that are stretched after spinning. Crystals are highly oriented, and material in the noncrystalline regions is also rather highly oriented. Orientation in the noncrystalline regions, as we have said, depends strongly on the processing history of the fiber. In the last decade or so, fiber producers have been placing emphasis on increasing the speed at which fibers are wound during spinning. Increasing the wind-up speed causes high orientation in the fiber prior to solidification by crystallization. The more orientation put into the fiber during spinning, the less fiber stretching after spinning is required; however, when PET is spun at high speed, a unique structure develops. The high winding speed causes high molecular alignment in the melt, so the molecules crystallize rapidly and substantially. As a result, the orientation in the noncrystalline regions is very low. The unique structure of this material is shown in Figure 3.28. The point of showing this particular structure is to show that fiber scientists can manipulate processing so as to achieve a wide range of structures. Do not become enamored with a single model and apply it to everything.

Specialty Fibers

The chemical structure of thermoplastic elastomers was described in the early part of this chapter. It is a combination of both the chemistry and the morphol-

FIGURE 3.25   Fibrils revealed by peeling drawn polyethylene (400 μm diameter fibrils).

FIGURE 3.26   Modified fringed micelle model.

FIGURE 3.27   Peterlin model of drawn fibers.

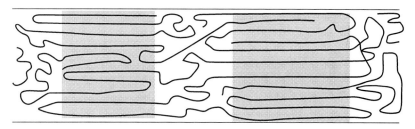

FIGURE 3.28   Structure of high-speed spun PET fiber.

ogy that give these block copolymers elastomeric character. The two blocks are chosen so as to be incompatible. That is, the chains of one block type seek out similar chains, so that the solidified polymer is not a microscopic mix of blocks, but rather a two-phase structure consisting of hard regions of block A and soft regions of block B, as shown in Figure 3.29. The continuous phase, the soft phase, needs to be elastic or fluidlike—a noncrystalline material with rubbery behavior. The discrete phase, the hard phase, needs to be glassy or semicrystalline. The hard segment anchors the ends of the soft segments, giving the molecules a memory of their location, acting much like crosslinks in vulcanized rubber. An example of a thermoplastic elastomer is the first one developed, which has polybutadiene soft segments (unvulcanized rubber) and polystyrene hard segments (glassy).

The structure of liquid crystalline polymers or that of the specially processed Spectra® 1000 PE discussed previously is unique. It is a near-ideal structure, with most of the molecules virtually completely stretched out. While this conformation is not desirable for a textile fiber, it is excellent for a fiber used for composite reinforcement. The structure is shown in Figure 3.30. It is often referred to as continuous crystal. A continuous crystal is relatively easy to achieve with liquid crystalline polymers, since in the melt or solution the molecules align in more or less parallel positions. To achieve the continuous crystal morphology with PE requires elaborate processing to avoid chain entanglements and chain folding. Gel spinning of ultra-high molecular weight (UHMW) polymer is used. In gel spinning, a small amount of UHMWPE is

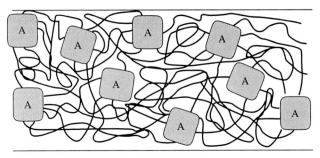

FIGURE 3.29 Structure of thermoplastic elastomers.

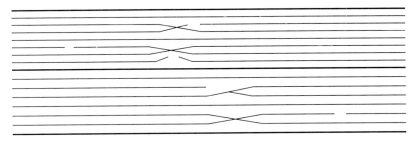

FIGURE 3.30 Model structure of Kevlar® or Spectra® 1000.

dissolved in a solvent. The viscous solution is wet spun, whereupon the spinning solvent is exchanged with a volatile nonsolvent, causing gelation of the PE. The gel is drawn at about 140 °C under conditions that give the polymer high orientation. The resulting fiber has high orientation, modulus, and strength.

The structure of glass in fiber form deserves a special note. Oxide glass, which is chiefly SiO_2, is noncrystalline. It is noncrystalline because it has the structure of a highly crosslinked polymer, as shown in Figure 3.31. Highly

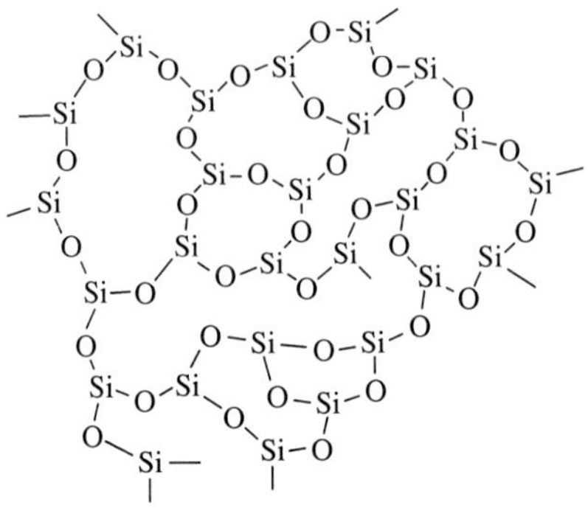

FIGURE 3.31 Schematic structure of silica glass.

crosslinked polymers cannot achieve orientation upon drawing. Hence, oxide glass fibers are single-phase—noncrystalline, glassy—and there is no molecular orientation.

3.4 Summary

In this chapter we have complemented our knowledge of the chemical structure of fibers with a knowledge of the physical structure of polymers and fibers. Amorphous materials are noncrystalline materials, materials in which the atoms are not in well-defined lattice positions. Noncrystalline solids may be glasses or rubbers, depending on whether they are brittle or not. Textile fibers are generally semicrystalline. Semicrystalline polymers can have good mechanical properties. Noncrystalline fibers are usually single-phase, in which case they may be transparent and perhaps useful for their optical properties.

The atoms in crystals are arranged on a lattice that repeats in space. Polymer crystals are more complex than most ionic or metallic crystals. The chains in polymer crystals are arranged linearly or helically. Large side groups favor helical conformation: Silk is stretched out; wool is helical. Polymers are typically not completely crystalline, as there is insufficient time for the molecules to untangle during crystallization and some polymer chains are difficult to pack efficiently. Hence, polymers are semicrystalline, containing regions of high order and regions of less order. X-ray diffraction is used to study crystal structures. Bragg's law is the defining equation for diffraction in materials. Crystallinity is the fraction of crystalline material in a sample. Fiber scientists often envision a two-phase model for the structure of polymers and fibers. The ease of use of a two-phase model—crystalline and noncrystallline—often overshadows the inaccuracies associated with the use of such an ideal model. Crystallinity cannot be determined with high accuracy. X-ray diffraction, density, and heat-of-fusion studies can each be used to estimate crystallinity.

A unit cell is the smallest volume of a material that has all the properties of the entire crystal. Unit cells can be used to calculate density and modulus of crystalline material, which has perfect molecular alignment.

Most polymer molecules coil on themselves in the melt. The object of synthetic fiber formation is to create a fiber with the molecules largely oriented along the direction of the fiber. This is achieved by stretching the polymer melt or solution prior to solidification, after solidification, or both. When polymer crystallizes without stress, it is called quiescent crystallization. When stress is present, crystallization occurs on the sides of highly oriented clusters of molecules called row structures. The overgrowth is lamellar.

Various models have been proposed to represent the structure of polymers and fibers. Recall the variety of structures for the various fibers we have discussed. Virtually any structure is possible for a man-made fiber, but most commercial organic textile fibers—natural or synthetic—can be loosely modeled using a modified oriented fringed micelle model. Crystallinity varies from fiber to fiber, so the volume fraction of various phases varies. Likewise, the crystallite sizes and the orientation of the noncrystalline regions vary. Essentially all oriented semicrystalline organic fibers are fibrillar. The smallest fibrils, often referred to as microfibrils, are alternating layers of crystalline and noncrystalline material along the fibril axis. Table 3.3 will serve as a useful reference for the microstructural features of natural and typical synthetic textile fibers.

TABLE 3.3 Fiber Structural Models

Fiber	Cross-Section	Structural Details
Cotton		Cellulosic, collapsed lumen, 100% disordered crystal, growth rings, helical-layered molecular structure, twisted ribbon, fibrillar nature, 4 cm staple, hygroscopic and absorbent, helix angle about 25°
Wool		Round and large, each fiber is a collection of many oriented cells, scales on surface, 10 cm long, very hygroscopic and absorbent, 40% crystalline, proteinaceous, 35° helix
Silk		60% crystalline, fibers 1 km long, hygroscopic and absorbent, textureless, proteinaceous, planar zigzag conformation
Rayon		Regenerated cellulose, wet spun, reduced molecular weight, highly absorbent, various mechanical properties possible
PET, nylon	any shape	Melt-spun synthetic polymer, slow to crystallize, apparel as well as industrial applications, nylons absorb water, PET does not, fibers usually roughly 40% crystalline, T_m(nylon 66, PET) ≈ 260 °C
PE, PP	any shape	Melt-spun synthetic olefin, crystallizes rapidly, highly crystalline, low density, good mechanical properties, any shape possible, PE is planar zigzag, PP is helical, T_m(PE) ≈ 130 °C, T_m(PP) ≈ 160 °C
Acrylic		Solution spun, degrades prior to melting, carbon fiber precursor, unusual order—liquid crystal-like

TABLE 3.3 *Continued*

Fiber	Cross-Section	Structural Details
Thermoplastic Elastomer		Highly extensible block copolymers that can be melt spun to any shape, poor strength, low modulus
Spectra® PBO, Kevlar®		High modulus, high-strength reinforcement fibers; PPTA is a dry-jet wet-spun nylon; Spectra® is ultra-high MW gel-spun PE, PBO is wet-spun azole; very high molecular orientation, fibrillar structure, continuous crystal
Oxide glass		Amorphous, unoriented fiber usually used for reinforcement or insulation, high modulus and strength, good thermal stability
Carbon fiber		Small-diameter distorted graphitic fiber, highly anisotropic, strong covalent bonds oriented axially, high modulus and high strength, used for reinforcement
Ceramic fiber, e.g., Al_2O_3		Isotropic crystalline granular reinforcement fiber with good thermal stability, high modulus, and fair strength

Special fibers we have described include liquid crystalline fibers, in which the molecules line up essentially parallel in domains. It is relatively simple to achieve high molecular alignment in liquid crystalline fibers and, hence, high mechanical properties. Silica glass is a noncrystalline fiber that has good optical and high mechanical properties. Thermoplastic elastomers are two or more phase materials. The hard phase anchors the molecules, which are otherwise relatively free to move with stress.

In this chapter we have studied and learned about the microstructure of various organic fibers, especially fibers used in textile applications. Although we have developed the knowledge of how to manipulate the structure and morphology of synthetic fibers within wide limits, we cannot begin to compete with the sophistication of nature. Consider, for example, the difficult task of making a hollow fiber that has a very thin wall and will not collapse when a lateral force is applied. Man-made fibers have molecules aligned along the fiber axis, giving high axial properties, but fibers easily split lengthwise in response to lateral forces. In addition, we do not have the technology to make thin-walled hollow fibers. Nature makes several, for example, the milkweed fiber shown in Figure 3.32, which is designed to float in air and carry the

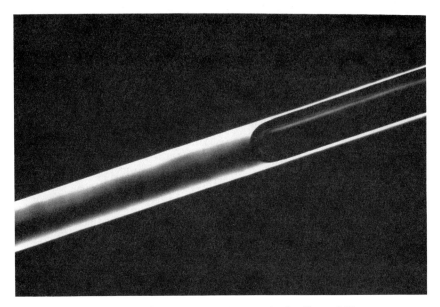

FIGURE 3.32 Optical photomicrograph of milkweed fiber.

milkweed seed. This fiber is about 24 μm in diameter; it has a wall thickness of about 1.2 μm and a helical conformation. The wall thickness can be seen as the thin, bright region in that section of the fiber that is filled with oil— the right side. Consider another natural phenomenon, spider webs. Spiders can synthesize and recycle many times crosslinked threads that are both very strong and highly extensible. We have not been able to copy this feat, despite years of trying.

References

D. C. Bassett. *Principles of Polymer Morphology*. Cambridge, MA: Cambridge University Press, 1981.

J. F. Drean, J. J. Patry, G. F. Lombard, and M. Weltrowski. "Mechanical Characterization and Behavior in Spinning Processing of Milkweed Fibers." *Textile Research Journal* 63 (1993), 443.

P. J. Flory. *Principles of Polymer Chemistry*. Ithaca, NY: Cornell University Press, 1953.

I. H. Hall, ed. *Structure of Crystalline Polymers*. New York: Elsevier, 1984.

E. J. Kramer. *Advances in Polymer Science* 52/53 (1983), 1.

W. E. Morton and J. W. S. Hearle. *Physical Properties of Textile Fibers*, 3rd ed. Manchester, England: The Textile Institute, 1993.

F. Rodriguez. *Principles of Polymer Systems*, 3rd ed. New York: Hemisphere, 1989.

S. L. Rosen. *Fundamental Principles of Polymeric Materials*. New York: Wiley-Interscience, 1982.

J. P. Schaffer, A. Saxena, S. D. Antolovich, T. H. Sanders, and S. B. Warner, *Materials Engineering*. New York: Times-Mirror Books, 1995.

A. Sharples. *Introduction to Polymer Crystallization*. London: Edward Arnold Publishers, 1966.

H. Todokoro. *Structure of Crystalline Polymers*. New York: Wiley-Interscience, 1979.

B. Wunderlich. *Macromolecular Physics: Crystal Melting*, V3. New York: Academic Press, 1980.

A. Ziabicki. *Fundamentals of Fiber Formation*. New York: Wiley-Interscience, 1976.

Problems

(1) Show with a sketch the structure of isotactic, atactic, and syndiotactic polystyrene. Polystyrene is a vinyl polymer with a benzene side group. Do you expect s-PS to be planar or helical?

(2) Given that the density of Spectra® 1000 PE fiber is 0.97 g/cm^3, estimate the crystallinity of the fiber.

(3) A rookie fiber scientist might notice that the fracture surface of most textile fibers observed in the microscope have a woody texture. What structural feature is responsible for this observation?

(4) Sketch the structure of a simple crystalline solid (such as Au) and show planes capable of X-ray diffraction.

(5) Give two reasons why polymers are less dense than most other materials.

(6) What three techniques can be used to assess the extent of crystallinity in a sample? Which of these techniques can be used to determine whether a polymer never before made is semicrystalline?

(7) How do wood pulp and cotton fibers differ?

(8) What are some of the morphological differences between silk and wool? Why is silk roughly 60 percent crystalline whereas wool only 40 percent crystalline?

(9) Sketch the physical structure of a Corriedale wool fiber.

(10) What is a unit cell? Why is it a useful concept?

(11) Calculate the fraction of volume filled by atoms in a crystal of Al, shown in Figure 3.2. Repeat the calculation for a central carbon bonded to four identical carbon neighbors. Assume carbon atoms touch. (The unit cell below is dotted.)

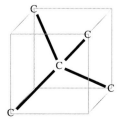

(12) Calculate the repeat spacing along the crystallographic c axis, the molecular chain axis, of polyethylene given the following X-ray information: A diffraction maximum occurs from the (002) plane at $2\theta = 74.3°$ using Cu radiation ($\lambda = 1.54$ Å). [The repeat spacing is double the separation of (002) planes.]

(13) Calculate the density of PE crystals from the unit cell information. Note that there are two mers per unit cell and all angles are 90°.

(14) When spinning a fiber, the wind-up speed is much greater than the extrusion speed, causing a diameter reduction of the extrudate from about 400 μm to about 20 μm and an increase in the molecular orientation. With such a high drawdown we anticipate very high molecular orientation, but it is not realized. Calculate the drawdown (fractional area reduction) and explain what causes the orientation to be less than anticipated during fiber spinning.

(15) Are all fibers two-phase materials? If not, then what?

(16) Under what conditions will polymer form spherulites? row-nucleated structures? Sketch the structure of a spherulite, a row-nucleated fiber, and a continuous crystal.

(17) Why do some solid polymers coil on themselves and others crystallize?

(18) Why do textile scientists try to stretch out and orient synthetic polymer molecules? Does nature do this in cotton or wool?

(19) Why are some polymers transparent?

(20) Estimate the crystallinity in a PET fiber with a density of 1.395 given that the extremes (0% and 100%) are 1.335 and 1.455 g/cm^3. Compare the value to that obtained from DSC, which shows that the heat of fusion is 60 J/g, given that the heat of fusion of the crystal is 140 J/g.

(21) How might you differentiate the following fibers?
(a) PET from nylon 66
(b) PE from PAN
(c) cotton from wool
(d) i-PP from a-PP
(e) PET from cotton
(f) rayon from cotton

(22) Nylon 6 may be either melt spun or solution spun. Why are all commercial samples of textile nylon 6 melt spun?

(23) Why is cotton one of the most used and therefore one of the most important textile fibers?

(24) Synthetic fibers are made in a continuous fashion, but often they are cut to a length of about 4 cm. Why are the continuous fibers converted to staple fibers?

(25) Sketch the structure of a glassy polymer using Figures 3.22 through 3.29 as models.

(26) A fiber is made by melt spinning isotactic polypropylene. Unlike most spinning operations, the fiber is not stretched during extrusion; rather, it is allowed to fall in air under its own weight. The process is called "free fall." Describe the structure of the fiber in as much detail as you are able.

(27) Many fibers are dyed using substances that form secondary bonds with the polymer molecules. Describe whether you expect acid dyes ($-SO_3H$ or their salts) to be effective at coloring cotton, wool, silk, and atactic polypropylene. Consider not only affinity, but also accessibility, and rank the fibers.

(28) Information is needed on the characteristics of nylon 6 crystals. To obtain the data, a fiber is prepared and analyzed. The DSC of the fiber of nylon 6 is as shown:

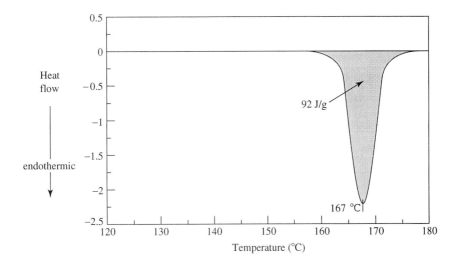

Optical microscopy shows the fiber is highly birefringent (oriented). X-ray diffraction of the fiber indicates the material is about 42% crystal. The density of amorphous nylon 6 is known to be 1.10 g/cm^3.

(a) Estimate the heat of fusion of pure crystal.

(b) Sketch a reasonable fiber X-ray diffraction pattern.

(c) Estimate the density of a nylon 6 crystal, given that the fiber density is 1.16 g/cm^3.

(29) Sketch the structure of stretched-out vulcanized rubber and stretched-out soft segments of thermoplastic elastomer. What might occur in these materials as they are highly stretched?

(30) What is the crystallinity of the PET textile fiber described by the data in Table 3.2? Estimate its heat of fusion.

(31) Calculate the Avrami parameters, k and n, for the isothermal DSC data shown here, obtained for thermotropic polyester with a composition similar to that of Vectra®. Approximate the crystal density to be insignificantly different from that of the noncrystalline material, so that weight fractions can be used in place of volume fractions.

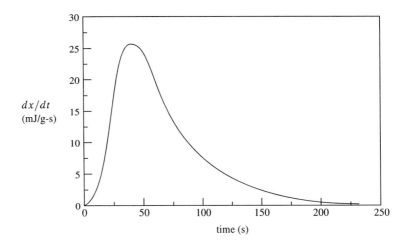

4

Staple Fiber Length and Sampling

In this chapter the importance of using proper sampling techniques is discussed and some of the effects of using improper sampling techniques are described. The fiber property that is contemplated is fiber length and its associated length distribution; however, other properties are also mentioned. This chapter is born out of the fact that two natural fibers, cotton and wool, are of significant commercial importance. These two fibers are harvested as staple fibers with a distribution in all physical and chemical properties, including fiber length. The attention directed to fiber length is warranted by the importance of fiber length in staple fiber processing and by the effect of fiber length on yarn strength. For a given type of natural fiber, fiber cross-sectional area, another important property addressed in the following chapter, often correlates with fiber length.

The importance of obtaining property values of a subsample that accurately reflect the entire sample's properties is of paramount importance and not a trivial issue. It extends well beyond the problem of determining mean fiber length or fiber length distribution in cotton. It is a major problem whenever a large sample is available and only relatively few tests are conducted. How does one ensure that the selected samples reflect the properties of the bulk sample? Think for a moment about the problem of determining the mean strength and strength distribution of a 100 kg bale of cotton fibers.

The focus of this chapter is not on statistics, but rather on teaching that statistics must be used throughout fiber science, especially in this area. I leave the statistics lessons to statisticians and focus on the fiber science.

The mean is a statistically simple value to compute, and it is used countless times in this chapter; however, the mean is a single value and actually gives little useful information unless the distribution is known. In practice, few distributions are determined. The reader is cautioned about the use of only the mean and standard deviation in characterizing a distribution. Consider cotton, for example. Shortly we will learn that long fibers generally yield higher strength yarns than do short fibers. Is mean fiber length important, or even indicative of yarn strength, or is the critical factor to assess the number or mass fraction of fibers longer than a given minimum?

4.1 Biased Sample Selection

Suppose you are charged with determining the average fiber length in a specific cotton yarn. You decide to select a yarn sample simply by stopping the yarn as it travels into a drafting (sometimes called drawing) zone. You select at this point because the yarn is conveniently firmly held between sets of pinch points. In commercial practice the exit set of rolls has a higher surface speed than the input rolls, forcing the drafting, as shown in Figure 4.1. This type of drawing, staple yarn drawing, is different from filament yarn drawing. In fiber drawing the diameter of each fiber is reduced, and work is done on the fiber. In yarn drawing or drafting, on the other hand, the fibers simply slide by one another, aligning fibers along the yarn length and reducing the number of fibers in the cross-section. The diameter of each fiber remains constant.

When the drafting process is halted, fiber may be combed away from one set of pinch points. Only those fibers securely held in the nip remain after combing. This sample is called a Wilkensen tuft. It consists of two beards, one protruding to the left and one to the right. The Wilkensen tuft is now removed from the machine, and each fiber length is measured. The histogram of fiber length is shown in Figure 4.2. The short fibers are to the left, the long fibers to the right. The height of each length (the ordinate) represents the fraction of sample with that length. The data may be plotted another way, as a survivor diagram, which some fiber scientists and statisticians find easier to use. Here the abscissa is fiber length, but the ordinate is the fraction of fibers with lengths exceeding that specified by the abscissa, as shown in Figure 4.3. Figure 4.3 has no more or less information than does Figure 4.2.

Let us return to the sampling technique. We have used just one location along the yarn to measure fiber length. When the yarn is short, this proce-

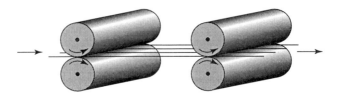

FIGURE 4.1   Schematic of drafting zone.

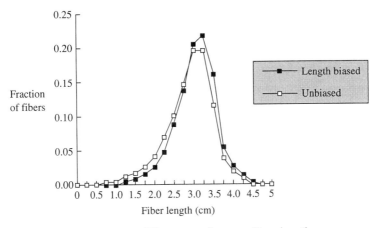

FIGURE 4.2 Histogram of cotton fiber length.

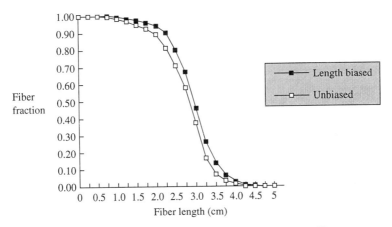

FIGURE 4.3 Survivor, or Baer, diagram of cotton fiber.

dure is satisfactory; however, when the yarn is long, we need to select various locations at random along the yarn. We will indicate how to do this in the last section on zoning and random sampling. In addition, other sampling techniques may be used. The Wilsensen tuft was mentioned only because it is a convienent way to obtain a sample in a processing operation.

The sample that we have just taken, the Wilkensen tuft, is not a random sample. We have not selected fibers at random. We have given preference according to fiber length. The longer the fiber, the more time it spends in the nip. Hence, the probability of selecting a fiber is directly proportional to its length. We have developed a histogram for a length-biased sample. The histogram that accurately reflects fiber length distribution in this sample will have fewer long fibers, as shown in Figures 4.2 and 4.3.

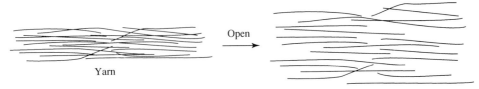

FIGURE 4.4 Opening a yarn.

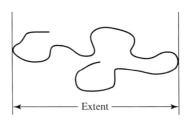

FIGURE 4.5 Extent biasing.

It is not harmful that we have selected a length-biased sample; however, it is unforgivable to *unknowingly* produce biased data. We can relate the mean length of the unbiased sample to that of the length-biased sample using the equation

$$L_{\text{biased}} = L_{\text{unbiased}} + \sigma^2_{\text{unbiased}}/L_{\text{unbiased}}$$

where σ is the standard deviation in length of the unbiased sample, so σ^2 is the variance. In fact, we can calculate all the important aspects of the unbiased sample from measurements on a biased sample. (For example, the unbiased fraction of fibers, f, of length, L, is $f = f_{\text{biased}}/L$.) We do need to know that we took a biased sample.

How can an unbiased sample be taken? There are several techniques in use. The important concept in the two described here is that all fibers have two ends, so end counting can produce unbiased results. One technique that is used with sliver (a collection of fiber the diameter of a rope) or yarn is to open the yarn normal to the long axis, as shown in Figure 4.4. A hoop is then placed over the opened yarn. Those fibers with one end in the hoop are counted once and those with both ends in the hoop are counted twice. Dye-sampling, another technique, was first developed for wool in carded web form, i.e., with fibers aligned in a 2-d fabric. A rectangle within the sample web is dyed. The fibers are then gently removed; those with one end dyed are counted once, and those with both ends dyed are counted twice.

In addition to length biasing, extent biasing can alternatively occur. The sampling just described is for fibers in slivers, yarns, or webs in which the fibers are highly extended. When selection of fibers is from a sample in which the fibers are not extended, then the sampling may be biased on the basis of the fiber extent, as shown in Figure 4.5. Sampling in this way produces extent biasing. A modern example of extent biasing is described in problem 2.

4.2 Fiber Length Measurement

When only a small sample is under consideration, fiber length may be measured one fiber at a time. Obviously, this may be a laborious, time-consuming approach. There is the added difficulty that cotton, wool, and other fibers have texture, which encourages the fibers to shrink from their stretched-out length. These problems can be overcome, such as by using an oiled glass, and manual measurement remains a viable technique for small samples.

When only mean fiber length is required, then a simple cut-and-weigh technique may be satisfactory. A sample of fibers is arranged with one of the ends of each fiber aligned with the others. A cut is then made a short fixed distance, L_1, from the aligned ends. The cut divides each fiber into two parts, as shown in Figure 4.6. The section containing the aligned ends is weighted as a group to obtain M_1. The second group is then weighed to determine M_2. A ratio is used to determine the mean L_2, $\langle L_2 \rangle$:

$$M_1/L_1 = M_2/\langle L_2 \rangle$$

The mean fiber length, $\langle L \rangle$, is the sum of the two lengths:

$$\langle L \rangle = L_1 + \langle L_2 \rangle$$

The technique is not accurate when fiber cross-sectional area tapers toward the fiber ends, nor is it capable of providing information on fiber length distribution.

Automated instrumentation is available for the analysis of the length and length distribution for samples with a large number of fibers. Typically a beard is used for the analysis. The optical density of the beard is determined as a function of position from the clamping line. The intensity of the light beam

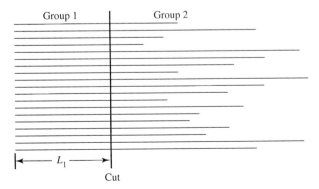

FIGURE 4.6   Cut and weigh technique to assess mean fiber length.

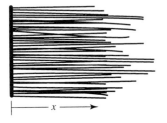 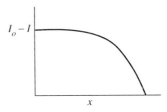

FIGURE 4.7 Beard and intensity scan through beard. (Intensity not transmitted is plotted on ordinate.)

transmitted through the beard increases as fibers fail to extend beyond their length, as shown in Figure 4.7. Note the similarity of the curve to a survivor diagram. Typical values of mean fiber lengths for various cotton, wool, and other fibers are shown in Table 4.1. In general, as fiber cross-section increases or the amount of texture decreases, longer staple lengths are required for processing and to achieve reasonable yarn strength. Cotton fibers have much texture, perhaps 40 twists of the collapsed fibers per cm length. The amount of texture in wool increases with decreasing fiber diameter, which is readily understood on the basis of the origin of the crimp.

TABLE 4.1 Typical Fiber Lengths

Fiber		Length (cm)
Cotton:	Bengals	1.2–1.5
	American Uplands	1.9–3.0
	Egyptian Uplands	2.7–3.2
	Sea Islands	3.2–3.8
Other Vegetable:	Flax	15–60
	Hemp	120–300
	Jute	150–360
Wood:	Softwood	0.2–0.7
	Hardwood	0.1–0.3
Wool:	Australian Merino (20 μm diameter)	6–7.5
	Corriedale (30 μm diameter)	7.5–9.0
	English Leicester (40 μm diameter)	25–35
Mohair:	Angora Goat	9–12
Silk:	Cultured Silkworm	150,000

Source: M. L. Joseph, *Introductory Textile Science*, 5th ed., New York: Holt, Rinehart and Winston, 1986; W. von Bergen, ed., *Wool Handbook*, V1, 3rd ed., New York: Interscience, 1963.

4.3 The Importance of Fiber Length

Fiber length is critical in processing of fibers and yarns and in the translation of fiber to yarn strength. Let us consider the former first and return to
the concept of drafting. Between the sets of nips, fibers slide past one another, reducing the cross-sectional area of the yarn. The process requires friction among fibers, but not so much that the fibers break rather than slip. An
additional requirement is that the fibers be sufficiently long that they are always under either the input or output nip. If the fibers are too short, then
the process may be difficult to control and lead to yarn with poor uniformity.
The short fibers may fall out of the yarn and collect under the drafting zone.
These noils, as they are called, are waste in the process of yarn manufacture.
In cotton yarns they account for perhaps 1 percent of the fiber. When they
get through drafting, however, they may adversely influence other processing
steps or yarn properties. The problem with noils, then, is not rooted in mean
fiber length, but rather in the *mass* fraction of fibers that are below a certain
minimum length. The short fibers are the most difficult to measure in length
and the ones most easily overlooked when sampling. Their numbers may increase when processing is so harsh as to break fibers.

Another reason to exclude short fibers from the finished product is that
fiber length influences yarn strength. We will use composite theory to show
in general terms why this is so. Consider a fiber that extends into and out of
a matrix, as shown in Figure 4.8. When the fiber does not adhere well to the
matrix, a pullout force causes the fiber to pull out of the matrix. When the
interfacial strength is high, on the other hand, the fiber breaks in response to
a pullout force. When the interfacial strength and fiber strength are balanced,
some fibers pull out and others break. In this case, the fiber pullout force,
which is related to fiber strength, is balanced by the retaining force, which is
related to interfacial strength. Mathematically,

$$F_{\text{pullout}} = \sigma_f A_f = \tau_i A_i = F_{\text{retain}}$$

where

σ is the fiber strength,

τ is the interfacial shear strength, which is a measure of the fiber-to-matrix
coupling, and

A's are the fiber cross-sectional and interfacial areas.

We now focus our attention on round fibers. The interfacial area is

$$A_i = 2\pi R L$$

and the cross-sectional area is

$$A_f = \pi R^2$$

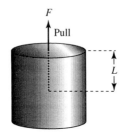

FIGURE 4.8 Single fiber pullout test.

where

    R is fiber radius, and

    L is fiber imbedment length in the matrix.

Substituting for the areas and solving for L gives

$$L = R\sigma_f / 2\tau_i$$

Note that the critical length depends on fiber radius, fiber strength, and the interfacial strength. Physically this means that when the imbedment length, L, equals the value above, called the critical value, half the fibers break and half pull out. If L exceeds this value, then the fiber breaks. Since fibers break rather than pull out, they are being used to their maximum stress-carrying capability. Conversely, short fibers cannot effectively reinforce a matrix.

At this point you may ask how this relates to yarns. Yarns are a collection of staple fibers (unless the yarn is a filament yarn of synthetic fibers). Each fiber can contribute to the yarn strength. Fiber strength is translated to the yarn by fiber-to-fiber friction, which takes some distance or length to develop, depending on the coefficient of friction, the contact area, and the lateral force, which are all largely determined by fiber texture and yarn twist. When a fiber is so short that stress cannot be transferred from other fibers to the short one, then the fiber is not an effective load-bearing element in the yarn. The fiber adds mass to the yarn, but it is not capable of carrying its share of the load. It is a defect. Twist may be added to the yarn, but twisting makes the yarn stiffer and harsher and reduces the potential strength because of the increased helix angle. (In Chapter 7 the relationship between strength and helix angle is derived: $\sigma \alpha \cos^2 \theta$.) Thus we see again that short fibers adversely affect at least one yarn property, strength, an important mechanical property that alone may determine the usefulness of a fiber.

4.4 Zoning and Random Sampling

Suppose you are presented with a trainload of staple fibers and you are asked to determine the fiber length distribution, strength distribution, or some other property histogram of the entire sample. Testing every fiber is out of the question. How might you proceed? We have already learned how to measure fiber

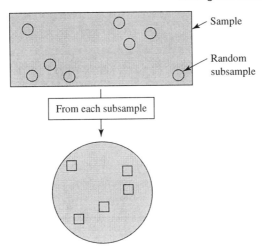

FIGURE 4.9 Schematic of zoning. From a sample, select at random several subsamples. In turn, select at random samples from each subsample. Continue the process until you have obtained a subsample that has few enough fibers that they can each be tested. The properties of the subsample should represent those of the entire sample.

length of a sample that we have selected. What remains is to select a subsample that accurately represents the properties of the entire bulk sample. How we proceed depends in part on the variability and scale of variability within the sample. If the sample is a 100 kg beam of industrial, continuous filament polyester, then it may be reasonable to conclude from knowledge of the formation process that the variability in strength from filament to filament will be about the same or slightly greater than that along the length of any filament. Hence, you *may* not need to break into the beam, and obtaining strength on a number of ends selected at random may be sufficient. The number of ends that you must test to ensure a meaningful value of strength depends on the sample size and on the variability in strength. You will learn to make those sorts of decisions in statistics class.

Suppose, on the other hand, you are given a bale of cotton fiber. You are warned that the fibers have not been blended in the bale and that there may be high variations in the quality of fiber from position to position in the bale. In this case you must select subsamples from various locations in the bale at random and test fibers from each of the subsamples. Zoning is the technique of using smaller and smaller subsamples until a reasonable number of samples is selected for testing, as illustrated in Figure 4.9. How do you know whether you have a representative sample? Again, you will learn more about this in statistics, but let it suffice to say at this point that replicate testing should give the same mean value and distribution of the property.

In Chapter 9 we will return to this subject in a much different form and discuss the effects of variability and describe continuous processes for measuring variability.

4.5 Summary

The focus of this chapter has been on tasks required of fiber scientists generally working with traditional, natural fibers. These skills and the understanding of these concepts must not be lost. Even if you do not work in a traditional fiber area, you need to appreciate the basic concepts discussed in this chapter. Moreover, statistics remains a vital element in the repertoire of skills available to a fiber scientist. You should be able to prepare a representative unbiased histogram from data you are given or from a truckload of fibers. This may include selecting samples using zoning, or whatever technique may be appropriate, and obtaining a meaningful fiber length distribution. Such data will inform you how much noil is in the sample and needs to be removed to ensure the smooth and efficient processing of staple fibers. Staple fiber length is important not only in yarn processing, but also in yarn properties, most notably yarn strength.

References

W. von Bergen, ed. *Wool Handbook*, V1, 3rd ed. New York: Interscience, 1963.

W. E. Morton and J. W. S. Hearle. *Physical Properties of Textile Fibers*, 3rd ed. Manchester, England: The Textile Institute, 1993.

G. W. Snedecor and W. G. Cochran. *Statistical Methods*, 8th ed. Ames, IA: Iowa State University, 1989.

Spin Off. D. Robson, ed. Published quarterly by Interweave Press, Loveland, CO.

Problems

(1) A sliver or yarn of mohair (similar to wool but no scales or crimp) is sampled by taking all fibers that pass through a plane normal to the yarn axis. Explain whether the sample taken is a random sample or somehow biased.

(2) Meltblown fibers are sized in a U.S. Patent (USP 4,622,259 to McAmish, Addy, and Lee, 1986) by noting the spacing of parallel wires that allows fibers to pass or not pass through during formation. Because of the high-velocity air stream carrying the fibers, the fibers can be modeled as random coils, as shown in the figure. Describe what the test really measures and whether it gives information on fiber length. The patent claims that as the wire frame is enlarged, then at some point no fibers will get hung up on the window. At this point the wire spacing provides a measure of the meltblown fiber length.

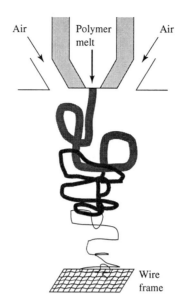

(**3**) What is a survivor diagram? a histogram? the dye test?

(**4**) What is a Wilkensen tuft? a beard? a noil? zoning?

(**5**) Suppose your job at Kimberly-Clark Corp. is to determine the mean diameter of round spunbond fibers in a given bonded nonwoven fabric 3 m wide and 1 km long. How might you do it? (Note that fibers are bonded together and cannot be individually removed from the fabric.) Be sure to explain how you ensure a random sample.

(**6**) Assume you accept a job at Amoco Performance Products upon graduation. You are asked to work in the carbon fiber testing lab. A customer brings you a 100 kg beam of continuous-filament yarn (1000 filaments) and asks you to determine the mean fiber strength. How will you proceed? Note that carbon fibers are brittle and the variability in strength is high.

(**7**) Suppose you work for Hoechst-Celanese Corp. in a continuous-filament polyester plant. There seems to be a chronic variation in fiber diameter. You hypothesize that the problem is a day/night problem, that is, that smaller fibers are made in the day and the larger ones at night. How would you proceed to prove your hypothesis?

(**8**) A yarn of Allied-Signal Corp. nylon is sectioned normal to the fiber axis so that the cross-section of each fiber is displayed. A line is drawn through a photomicrograph of the fiber ends, as indicated here:

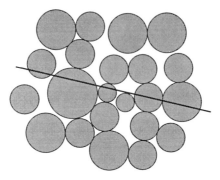

In general, do the fibers that intersect the line (or any other straight line) give a random sample? If not, then suggest a technique that will give a random sample.

(9) The fibers of a Shaw Ind. wool carpet yarn are carefully separated without disturbing the relative position of the fibers along the yarn axis. A hoop is placed over a section of the separated yarn. The lengths of the fibers with one or more ends in the circle are measured and the results are plotted in a histogram. (Each fiber with both ends in the hoop is counted twice.)

    (a) Sketch the histogram and put realistic numbers on the abscissa and ordinate for Corriedale wool.

    (b) Describe (and show on the histogram) how the data might differ from unbiased data.

(10) You are given a sample of wood pulp fibers. They are quite short, roughly 2 mm long. You use an imaging technique to obtain a photomicrograph of a representative subsample of the fibers. Can you get a representative histogram of fiber length by drawing a line at random, such as the one shown, and measuring the length of each fiber that crosses the line? (You do this many times on many subsamples.) You must state your reasons.

(11) It is customary to shear Merino sheep more frequently than other types of sheep, because the finer wools make as strong a yarn as do the coarser wools even with shorter staple fiber length. Explain using words and mathematics why this is true.

5

Fiber Cross-Section and Linear Density

Two of the most important physical properties of fibers are cross-sectional shape and cross-sectional area, a fact that has been recognized for centuries. Fibers with small cross-sectional area lead to more uniform and lightweight yarns than do coarse fibers. Also, fibers with small cross-sectional area are the most valued for soft fabrics, such as fabrics that are in contact with skin, and fabrics that cover well. Fibers with large cross-sectional area are valued in applications where stiffness is paramount, such as in carpets and suit coats. In addition to influencing bending and torsional stiffness, cross-sectional area and shape largely determine the optical properties of fibers. In some applications fibers with high specular reflectivity are used to give a glossy or glittery effect. In other applications low specular reflectivity is required to give a dull or subdued appearance. Fiber cross-section also influences yarn strength. The strength of a twisted yarn is determined by the ability of each fiber to transmit stress via surface interactions to adjacent fibers. Fibers with large surface area-to-volume ratios (small fibers) are better able to transmit stresses. Finally, liquids and vapors diffuse into small fibers more quickly than into large fibers. Assemblies of small fibers can hold fluids more tenaciously than can assemblies of large fibers; olefinic microfibers, for example, are used to absorb oil spills and wood pulp fibers have historically been used to absorb in infant diapers.

The focus of this chapter is characterization of fiber cross-sections. The influence of fiber cross-sectional area and shape on theoretical and actual fiber physical properties is also addressed, but much more on this subject follows in the remainder of the text.

5.1 Linear Density

Linear density is the mass per unit length of a long, slender object, in this case, a fiber. It is a convenient way to characterize a fiber, since the cross-sectional shape of fibers varies from fiber to fiber and often along a single staple fiber's length. In addition, we find it desirable to express the fiber's strength and stiffness in units that facilitate comparisons among different fibers with different densities. Hence, we will use the units of specific strength, or tenacity, and specific modulus, which are simply load-to-fail and load per unit of normalized extension each divided by linear density. Let us begin with a discussion of fiber linear density, move on to yarns, and then consider fiber cross-sectional shapes. The next topic addressed is the various techniques that may be used to measure linear density. We conclude with a discussion of cotton fiber maturity, which is a measure of the fiber's wall thickness.

5.1.1 Fiber Linear Density

For centuries fibers have been classified according to their cross-sectional area. Wool, for example, is classified using a numerical scale. Fine wool, such as Merino, can be as high as the 80's and coarse wool, such as Lincoln, is in the 36's. The classification is determined by a panel of experts who subjectively assess the quality of fibers that differ only in source. The values given by the panel of experts correlate directly with fiber diameter. The fineness of wool is a direct function of the sheep's heritage and, hence, their bloodlines.

Fiber scientists prefer to use linear density to characterize the cross-sectional area of fiber. Linear density is the mass per unit length of fiber. Historically the linear density has been reported in units of denier, which is the mass of fiber in grams per 9000 m of length. Denier is a convenient unit, since most individual fibers are in the 1- to 10-denier range. The higher the denier, the larger the cross-sectional area. Fiber linear density is also commonly reported in units of tex or decitex, which are the mass of fiber in grams per 1000 or 10,000 m. Note that fiber denier equals $9 \times$ tex of the fiber or $0.9 \times$ decitex. (The conversion factor is 9 denier/tex.) Typical values of fiber denier are provided in Table 5.1. Note the coefficient of variation, CV, equals the standard deviation divided by the mean.

Natural and synthetic fibers that are used against the skin typically have low denier. Fibers with high cross-sectional area are used in applications requiring stiffness. The cross-sectional shape and area of man-made fibers are manipulated to give the desired result, tailored to the application. One area of current research and development is the formation of very small-denier fiber. Uniform fibers with denier less than 1.0, so-called microfibers, have been historically difficult and costly to achieve; however, consumers are purchasing the more expensive soft fabrics that contain fine fibers.

Microfibers can be made in several ways. The most straightforward way is to extrude as usual using as small a spinneret hole as possible and achieve

TABLE 5.1 Typical Fiber Deniers

Fiber	Denier	Within Sample CV
Wool		0.30
Minimum (Merino)	4	0.14
Maximum (carpet)	20	
Vicuna	1.7	0.15
Chinese cashmere	1.8	0.19
Silk	1.0	
Cotton		
Minimum (St. Vincent Sea Island)	1	
Maximum (native Indian)	3	
Vicara	2.5	0.12
Synthetic		
Minimum (melt blown)	0.01	> 1.0
Maximum (monofilament)	10,000	
Typical apparel	0.9–3	0.12
Typical industrial	3	0.06
Typical carpet	6–20	0.12

Sources: W. von Bergen, ed., *Wool Handbook*, V1, 3rd ed., New York: Interscience, 1963; W. E. Morton and J. W. S. Hearle, *Physical Properties of Textile Fibers*, 3rd ed., Manchester, England: The Textile Institute, 1993; author's own data.

as much area reduction through drawdown as possible. By using good clean polymer and well-controlled process conditions, the spinneret plug rate and fiber break rate can be minimized. With current technology, fiber denier is limited to about 0.9 for PET and 0.6 for PP. Another way to obtain smaller microfibers is to spin two polymers together, so-called bicomponent spinning, using the sea-island morphology. When viewed in the cross-section, the polymer of interest consists of islands in a sea of the second polymer.

After processing, the sea is dissolved with an appropriate solvent, leaving the microfibers, which may be quite small. Another technology for making microfibers is fibrillation. In this technique a fiber or film is highly oriented and then broken down into its component fibrils using mechanical, chemical, ultrasonic, or other processing.

5.1.2 Yarn Linear Density

The linear density of yarns may be characterized in terms of denier or tex, which are referred to as direct systems, or in terms of length per unit mass or reciprocal denier, which are referred to as indirect systems. Indirect count systems reflect the history of textiles, going back to old England, where skeins of yarn were wound on a device called a weasel, which had a wooden stick that popped up when the appropriate number of turns of yarn had been wound on the reel. (Hence, the children's nursery rhyme chorus, "Pop goes the weasel.") These skeins or hanks were placed on a balance opposite a 1 lb mass. Although the King of England defined a pound, he did not define a hank. Hence, there are nearly 200 yarn-count systems with differing length hanks, but fortunately, only three account for all the indirect-count systems in the United States: cotton, metric, and woolen. Cotton count is by far the most commonly used system. An indirect system may be defined as the number of hanks required to sum to a unit mass, as shown in Table 5.2. Conversion from one system to another requires remembering only that the count times the hank length in each indirect system (except the metric) is the yards of yarn per pound of yarn. Note also that cotton count $= 2/3 \times$ worsted count, since $560/840 = 2/3$, and metric count $= 1.69 \times$ cotton count. Cotton count $\times$ denier $= 5315$, a dimensionless constant.

The notation used to express cotton count is the number of hanks followed by s, such as 10s or 45s. A single yarn is referred to as a singles yarn, say a 10s. Two 10s singles yarns plied together are 10s/2, which has about the net count of a 5s yarn. When three worsted wool 12s yarns are plied together, the result is a 3/12s yarn. The concept is the same, the notation a little different.

At this point you should be able to calculate denier or count from fiber cross-sectional dimensions and density. For example, the denier of a round fiber can be calculated from first principles. From the definition of denier:

$$\text{denier} = \text{volume of 9000 m} \times \text{density}$$
$$= A \times 900{,}000 \text{ cm} \times \rho$$
$$= \pi R^2 \times \rho \times 9 \times 10^5 \qquad [R \text{ in cm}]$$
$$= \pi R^2 \times \rho \times 9 \times 10^{-3}$$

with R in μm and ρ in g/cm^3. Thus,

$$R \ (\mu\text{m}) = [1000 \times \text{denier}/(\pi \times 9 \times \rho) \ (\text{g/cm}^3)]^{1/2}$$

You should also be able to convert from one count system to another and from denier to count. Several end-of-chapter problems test your knowledge in this area.

TABLE 5.2   Indirect Count Systems

Name of System	Hank Length	Unit Mass
Cotton, Ne	840 yds	1 lb
Worsted, Nw	560 yds	1 lb
Woolen	1600 yds	1 lb
Linen (lea)	300 yds	1 lb
Metric, Nm	1000 m	1 kg

5.2 Fiber Cross-Sectional Shape and Surface Area

The shape of a fiber's cross-section is also important in many applications. Consider, for example, bending rigidity. Of the solid fibers, those with round cross-sections offer a high resistance to bending and, hence, the fibers are stiff. Uncollapsed hollow fibers can be even stiffer. Fibers with ribbonlike cross-sections, on the other hand, such as cotton or wood pulp fiber, offer the least resistance to bending. These results are derived using simple elasticity theory in Chapter 8, where all the factors affecting fiber stiffness are presented.

Surface area is the fiber–air interfacial area of a fiber. It is customarily reported as specific surface area, which is surface area normalized to either fiber mass or volume. Specific surface area is not another variable, but once the denier and cross-section are set, the specific surface area may be calculated. Consider, for example, a round fiber. The surface area of the fiber per unit length is $2\pi R$, and the fiber volume is πR^2, so that the specific surface area reported on a volume basis is precisely $2/R$. The specific surface area reported on a mass basis is simply $2/R\rho$. Similar calculations may be conducted on any cross-sectional shape. A round cross-section gives the minimum surface area per unit volume for a continuous fiber. The larger the radius of the fiber, the smaller the specific surface area.

In Chapter 3 we discussed the shapes and manufacture of various man-made fibers. Examples of bilobal, trilobal, hollow, and other shapes can be found. A similar technology is bicomponent fiber spinning. When two polymers are extruded together in fiber form, a bicomponent fiber is formed. The two components can be arranged in symmetrical or asymmetrical sheath-core or side-by-side:

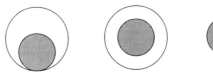

Bicomponent fibers are an attempt to get the best of both polymers in a single fiber. Bicomponent fibers are commonly used in nonwoven technology, where the lower melting surface or side facilitates bonding and the more thermally stable component retains its molecular orientation and mechanical properties through bonding. Asymmetric bicomponent fibers are also self crimping, with sections along the length each acting like bicomponent strips.

At this critical point you may wish to return to the last table in Chapter 3, which is a summary of the important physical characteristics of various fibers. Armed with these data, your knowledge of fiber chemistry, and the importance of fiber cross-section, you are now able to anticipate fiber characteristics and, hence, explore some potential applications of fibers on the basis of fiber chemistry, physical structure, and cross-sectional area and shape.

5.3 Variability in Fiber Denier

The denier of natural fibers varies both randomly and systematically. The rate of growth of natural fibers depends on a multitude of genetic and environmental factors. Growth factors may change from day to day, giving rise to brief or long-term fluctuations in denier. Consider, for example, a growing wool fiber. Should the sheep have a poor diet or be subjected to stress for a day or week, then the wool fiber will suffer, causing a thin and weak cross-section. This principle may indeed be used to advantage. Sheep can be injected with an appropriate (experimental) drug that weakens wool fibers for a brief period of time. A week or so later the entire fleece may be easily removed without shearing, as shown in Figure 5.1. Growing cotton is also subjected to fluctuations in weather and nutrition, leading to the occurrence of a variety of defects, or imperfections, in the fiber. The defects cause variations in fiber denier, strength, stiffness, dyeing characteristics, and so on.

Natural fiber denier can also vary systematically. Consider, once again for example, a wool fiber. The first fiber grown on a lamb is much finer than that on an aging sheep. Virgin wool is more valuable in textile applications, since the fiber is softer due to the taper at one end, as shown in Figure 5.2.

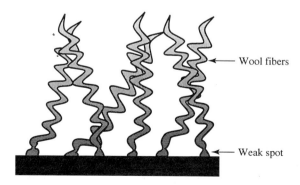

FIGURE 5.1   Wool being removed after injection of drugs.

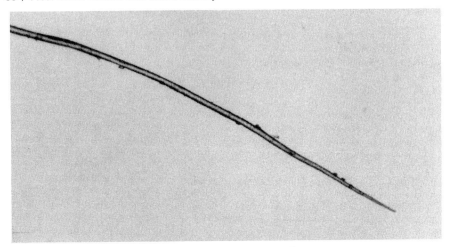

FIGURE 5.2 End of a wool fiber.

Cotton fibers also decrease in cross-section as the natural ends of the fiber are approached.

Man-made fibers may be formed with little variation in cross-sectional area. In optical waveguides it is important that fiber be invariant in diameter along the length of the waveguide. Hence, the fiber diameter is measured during formation, and a feedback loop is provided to change process conditions should the fiber diameter begin to deviate from specification. A schematic of optical fiber formation is shown in Figure 5.3. Textile fibers do not have such demanding tasks as do optical fibers. Hence, fiber diameter need not be and typically is not closely controlled. Industrial fibers require greater strength and uniformity than textile fibers, so their diameter is better controlled. Textile fiber spinning, indeed, utilizes the economy of scale. Thousands of fibers are spun at one time. Industrial fibers are spun several hundred at a time. Optical fibers are spun one at a time. Thus, the denier of nominally identical textile fibers varies from fiber to fiber as well as along a fiber, as indicated in Table 5.1.

The variability in denier has consequences in virtually all properties—dyeing, mechanical properties, optical properties, etc. The reason that fiber denier variability affects so many properties is that denier variability in synthetic fibers is an indicator of other variabilities, most notably molecular orientation. Diffusion rate, crystallinity, modulus, strength, density, refractive index, and a host of other physical properties depend on molecular orientation. Some of the consequences of the variability in denier and structure and, hence, mechanical properties, are the subject of Chapter 9. Techniques to measure variability in cross-section need to be considered first, and they are discussed in the following sections. Continuous, or so-called on-line, techniques for monitoring denier variability are discussed in the final section of Chapter 9.

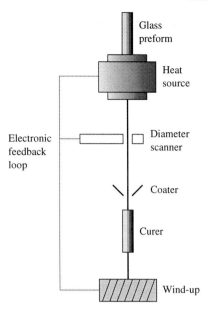

FIGURE 5.3   Formation of optical fiber showing feedback loop.

5.4 Measurement of Denier

There are a number of ways that denier can be measured. The best technique depends to a large extent on the characteristics of the fiber. No one technique is best for all samples. Thus, it is important to understand and be prepared to use all the techniques.

5.4.1 Direct Weighing—Whole and Cut Middle

The most direct and often the best technique for determining denier is simply to measure the length and corresponding mass of the fiber or a representative sample of interest. Denier is the mass in grams of 9000 m. The direct-weighing technique provides a meaningful measure of denier for a uniform fiber or group of fibers. When fiber denier varies along the length of a fiber or when a sampling of fibers with different deniers is included, the technique provides only an average value.

Suppose you have a wool fiber tuft and you need to determine denier. You may proceed much the same way as you did in Chapter 4, Figure 4.6. Align the fibers so they are substantially parallel, place a ruler normal to the fibers, and cut on both sides of the ruler. Multiply the cut length, L_1, with the number of fibers, n, to determine the total length. Weigh to determine the mass, M_1, in grams of the cut section of fibers. Divide the total length into the total mass and normalize to 9000 m. This technique provides an accurate measure of denier so long as the fibers do not taper toward the ends. Many natural fibers do taper at the ends, often leading to errors on the order of 15 percent.

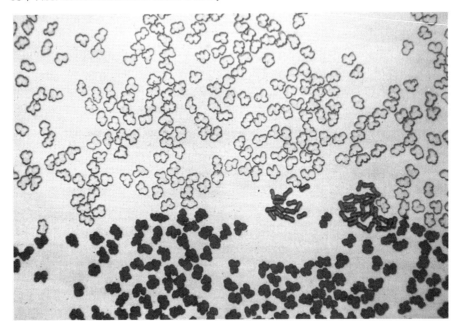

FIGURE 5.4   Cross-sections of fibers prepared by wet spinning.

Consider how you might determine the denier of a single wood pulp fiber that is only a few mm long. Direct weighing may not be the best choice, since the fiber will weigh only a few micrograms.

5.4.2 Microscopy

Optical microscopy may be used to determine both a fiber's cross-sectional shape and cross-sectional area. To convert cross-sectional area to denier requires a knowledge of fiber density. The microscope is used to obtain images of cross-sections of a representative sample. Either transmitted or reflected light microscopy may be used; however, sample preparation differs for the two techniques. In fact, a variety of techniques may be used to prepare samples for optical microscopy. One of the best procedures is to align fibers and mount them in an embedding medium of comparable hardness. Then, 5- to 15-μm-thick cross-sections are taken using a microtome. The sections are observed in transmitted light microscopy. Care must be taken to ensure that the optical path is along the fiber axis. Otherwise, distortion of the cross-section will produce erroneously high cross-sectional areas, unless sophisticated techniques using a universal stage are employed. Fiber cross-sections, as shown in Figure 5.4, can be analyzed manually using stereological techniques, planimetry, or computer graphic instrumentation. A crude observation of fiber cross-section may be conducted with the use of a Shirley Plate.

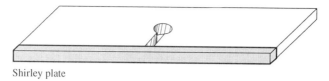

Shirley plate

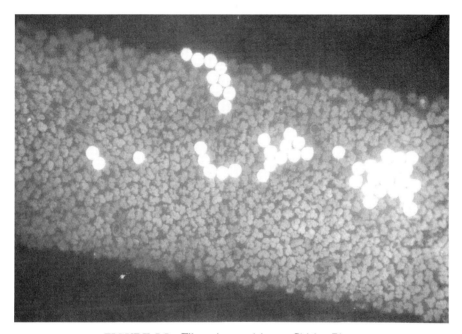

FIGURE 5.5   Fibers jammed into a Shirley Plate.

In this technique the sample of interest is mixed with standard fibers to form a uniaxially oriented yarn. The yarn is wedged into a narrow groove in a thin metal plate, shown in Figure 5.5. The fibers are held into position by a press fit cap (shaded). A sharp razor blade is slid over the two surfaces of the plate, leaving behind a section of fiber. The plate is placed on the stage of a microscope and the fibers are observed on-end in transmitted light. This technique may not be highly accurate because the long fiber length precludes formation of a sharp microscopic image.

In recent years transmission and scanning electron microscopy have been used to assess fiber cross-sections. While these techniques can be used, they are generally not necessary, and they are no better than optical techniques, except for very small fibers (i.e., $< 10 \ \mu$m diameter). When using an electron microscope, you must be sure that you take representative measurements and that only fibers in-focus are used. Out-of-focus fibers appear larger than they are. Figure 5.6 is a sample of a meltblown PP fiber web that was used to assess fiber diameter. A photomicrograph was taken at low magnification. A straight line was drawn through the micrograph. Every fiber that crossed the

FIGURE 5.6   SEM of a meltblown PP web.

line was examined in the SEM (scanning electron microscope). Each fiber was individually focused. Six low-magnification areas were used in total for the analysis. More than 100 fibers were analyzed. The resulting histogram of fiber diameters is shown in Figure 5.7. Meltblown PP fibers are round and small, with diameters typically in the range of 2 to 6 μm. Although only a fraction of the fibers measured is included in this histogram, it can be shown that the log of the fiber diameter is distributed normally. Given that meltblown fiber diameters follow a log normal distribution, only a few fibers in a new sample need be measured to provide a useful value for average or modal fiber diameter.

I do not mean to dismiss the use of SEM to obtain impressive images of fiber cross-sectional shape. The photomicrograph shown in Figure 5.8 is an SEM of trilobal nylon carpet fiber with a round conducting fiber. Unfortunately, it can be difficult to obtain meaningful quantitative information on fiber cross-section from such images.

5.4.3 Vibrascopy

One of the oldest, most elegant, and most precise techniques for determining denier is with a vibrascope. A single fiber is hung in a vibrascope and loaded with a mass, M, to produce tension, Mg, where g is the gravitational acceleration. A loudspeaker or transducer is used to excite the fiber. The frequency of excitation or the sample length is changed until the fundamental frequency is realized by noting when the center of the fiber attains maximum amplitude, as indicated in Figure 5.9. Automated instruments determine the fundamental frequency using capacitance or other electronic measurements. Manual instruments rely on telescopic observation of the fiber at its midpoint.

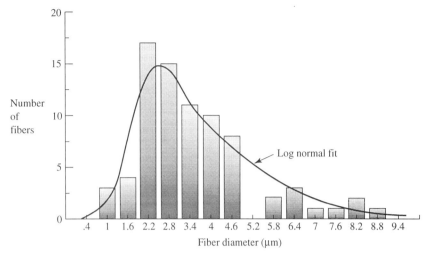

FIGURE 5.7 Histogram of meltblown PP fiber diameters.

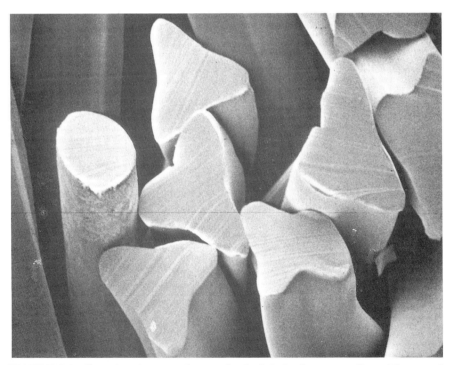

FIGURE 5.8 Scanning electron micrograph of trilobal nylon carpet fiber. (The round fiber is a conductor to dissipate static charge.)

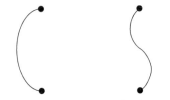

Fundamental frequency   *Higher frequency*

FIGURE 5.9   Two of many possible fiber profiles in a vibrascope.

A vibrascope works on the principle of a vibrating string, as sketched in Figure 5.9. The fundamental frequency of vibration, f, of a string or fiber is a function of the tension in the fiber, T, the mass per unit length, m_l, and the test length, l:

$$f = (1/2l)(T/m_l)^{1/2}$$

Solving for m,

$$m_l = T(1/2lf)^2$$

Note that the units of frequency are cycles/sec. Sec also enters the equation on the right through acceleration due to gravity (T).

5.4.4 Techniques Based on Air Flow

Laminar flow of fluids such as air through porous media is a well-understood phenomenon. The rate of flow through a porous plug, as sketched in Figure 5.10, is described by the Kozeny equation, which may be expressed as

$$Q = (1/k) \times (A\Delta p/S^2 \eta L) \times [\epsilon^3/(1-\epsilon)^2]$$

where

Q = volumetric flow rate through the plug,
k = geometric constant related to pore shape,
A = cross-sectional area of plug,
Δp = pressure drop across the plug,
S = specific surface area (area/volume) of plug material,
η = viscosity of fluid, usually air at room temperature,
L = length of plug, and
ϵ = porosity of plug.

To assess the denier of fibers, place a standard mass of fiber in the plug cavity, place the plug in the instrument at hand, increase the input air pressure until the standard pressure drop is achieved, and measure the flow rate. We can express Kozeny's equation in a different fashion by noting that porosity is the mass of the unfilled space if it were filled with polymer, normalized to

FIGURE 5.10   Air flow measurement of fiber denier.

the mass of the total space were it also completely filled with polymer:

$$\epsilon = (\rho AL - m)/\rho AL$$

where

ρ = density of the fiber, and

m = mass of the plug.

Hence,

$$Q = [\Delta p(\rho AL - m)^3]/k\eta m^2 \rho S^2 L^2$$

or

$$Q \propto 1/kS^2$$

where k is a geometric parameter that must be determined experimentally. It depends on the fiber and the amount of compaction in the plug. Once k is determined for a sample, it is a relatively simple matter to continue to use experimental conditions that allow the assumption that k is invariant for that type of fiber. Hence,

$$Q \propto 1/S^2$$

Since only two of the parameters—surface area, denier, and specific volume—are independent, once the mass and surface area are known, the denier can be determined.

Not all commercial instruments fix Δp and measure Q; rather, some instruments fix Q and determine Δp. Either technique is satisfactory. Modern instruments are completely automatic, providing an average sample denier that is as accurate as the k input into the calculation. You will be familiarized with this sort of testing in one of your laboratory courses.

5.5 Maturity of Cotton

As cotton fiber grows, the wall of the hollow cell thickens. Maturity is a measure of the extent that the walls have thickened. Immature, or "dead," cotton is unresilient and flimsy. The chief reason that immature cotton is undesir-

Immature fiber   *Mature fiber*   *Overmature fiber*

FIGURE 5.11   Maturity of cotton fiber.

able is that immature cotton fibers tend to form small, tightly wound balls of fiber called neps, which appear as unsightly specks in fabrics. Neps develop during processing as fibers rub one another and pass over surfaces. On the other hand, overmature cotton is to be avoided, since it is stiff and boardy. The lumen has been virtually closed by growth of the cell walls, as illustrated in Figure 5.11, and the virtually round cross-section has a high resistance to bending. We will measure maturity, θ, by taking the ratio of the cell wall area of the cotton fiber, A, to that of a circle with a perimeter equal to that of the cotton fiber, A_o:

$$\theta = A/A_o$$

For the overmature fiber in Figure 5.11, $\theta = 1.0$. A target range for maturity is between 0.8 and 0.9. Next we will discuss various techniques for determining maturity.

5.5.1 Measurement of Maturity Using Optical Microscopy

The optical microscope may be used in any of several ways to assess cotton fiber maturity. Cross-sections of fibers can be prepared by mounting the fibers, sectioning, and examining the cross-sections in transmitted light. Photomicrographs or digitized images may be analyzed using image analysis techniques to give θ as just defined.

Swelling experiments have also been used to assess fiber maturity. Samples of cotton fiber are placed between slide and cover slip and observed in transmitted light. A drop of 18 percent sodium hydroxide aqueous solution is placed on the slide and flows to immerse the sample. As the fluid touches the fiber, the fiber responds immediately by swelling. If the lumen is virtually eliminated by the swelling, then the fibers are normal, as shown in Figure 5.12. If the lumen in the swollen fiber is one third the fiber diameter, then the fibers are dead or immature.

A general assessment of the maturity of cotton fibers can be made by examining them in polarized light. Fibers are laid on a slide, immersed in an inert refractive index fluid, and covered with a cover slip. The sample is placed under crossed polars and a first-order red plate is inserted in the accessory slot in the microscope. Since all the fibers have about the same molecular orientation, the variations in color from fiber to fiber are due to two factors: (1) the position of the fiber relative to that of the polarizer and (2) the wall thickness.

FIGURE 5.12   Fiber swelling with NaOH to assess maturity.

Fibers that are yellow-green and remain this color with rotation of the stage are fully mature. Fibers that are blue-green or blue and yellow with rotation of the stage are partially mature. Fibers that are deep blue, indigo, or orange at stage rotation are immature.

5.5.2 Differential Dyeing as a Test for Maturity

The principle of the dye test is that immature fibers have a higher specific surface area than do mature fibers. Hence, immature fibers accept more dye in a short time than do mature fibers, but dye penetrates deep into a mature fiber when the fiber is exposed to the dye bath for a long time. Hence, dye can be washed out more rapidly from immature fibers than from mature fibers.

A 3 g sample of cotton fiber is immersed into a boiling dye bath of Diphenyl Fast Red and Chlorantine Fast Green. In 15 minutes, 4 percent by weight of fiber of sodium chloride is added. After 15 additional minutes, another 4 percent salt is added. After 45 minutes in the bath, the fiber is removed and rinsed three times in distilled water. After draining, the fibers are vigorously stirred in boiling water for 30 seconds. The centrifuged fiber is rinsed in cold water, dried, ground to a powder, and pressed into a pad. The pads are compared with standards. Mature samples are red, and immature samples are green.

5.5.3 Other Techniques for Determining Cotton Maturity

Air-flow techniques may be used to investigate fiber maturity. The same apparatus used to measure specific surface area or denier is employed. The principle is that immature fibers have a higher specific surface area than do mature fibers. A complication that may occur is that the pressure of the air flow may compress immature fibers, causing the plug to compress, block air passage, and give inaccurate results. Standards for this technique have yet to be developed.

Other techniques are constantly being evaluated, such as near infrared reflection spectroscopy and microscopic techniques. (See, for example, E. K. Boylston, D. P. Thibodeaux, and J. P. Evans, *Textile Research Journal* 63 (1993), 80.)

5.6 Summary

The cross-sectional area and shape of fibers have been considered in this chapter. They are important characteristics of fibers, largely determining the fiber's adequacy in potential applications that require optimal mechanical, optical, and tactile aesthetic properties, such as comfort. Both direct (mass per unit length) and indirect (length per unit mass) systems of characterizing linear density were discussed. In the indirect system, the length per pound of sample is used. Length is expressed as a number of hanks. In the direct system, the unit is grams of fiber per set length. The set length is 9000 m in the denier system and 1000 m in the tex system. The fiber count and denier are independent of fiber cross-sectional shape; however, cross-sectional shape is nonetheless important. Fiber denier or count may be determined using a variety of techniques, the four most important being direct weighing; use of a vibrascope; use of optical microscopy and known densities; and, for large samples, use of an air-flow apparatus. Cotton fiber maturity is a measure of the cotton cell wall thickness. Immature and overmature fibers are to be avoided. The former is flimsy and the latter is bristly. Cotton maturity can be measured using a number of techniques, but perhaps the best is to prepare fiber cross-sections for microscopy and use image analysis techniques to assess maturity.

References

J. Bear. *Dynamics of Fluids in Porous Media.* Toronto: American Elsevier, 1972.

W. von Bergen, ed. *Wool Handbook*, V1, 3rd ed. New York: Interscience, 1963.

N. Heimbold and J. Betts, eds. "Microdeniers for Apparel." *Inside Textiles* 10, 2 (1990), 1–2.

W. E. Morton and J. W. S. Hearle. *Physical Properties of Textile Fibers*, 3rd ed. Manchester, England: The Textile Institute, 1993.

D. O. Taurat. "Spinning of Conjugate Fibers and Filaments." *International Fibers Journal* (May 1988), 24–32.

S. B. Warner, C. A. Perkins, and A. S. Abhiraman. "Meltblown Polypropylene Fibers." *INDA Journal of Nonwovens Research* 2 (1990), 33–40.

Problems

(1) 167,160 yards of cotton yarn weigh 5.0 lbs. What is the cotton count and denier?

(2) How much will 40,000 yards of 5s cotton count yarn weigh?

(3) Calculate the mass in ounces of 9000 yards of woolen yarn measured as a 4s in the woolen run system.

(4) Convert 2.0 dpf Vectra® fiber to decitex.

(5) 220 denier PET yarn has what metric count?

(6) Suppose you are considering changing from 106.3 denier nylon yarn to cotton yarn for a particular application. What cotton count has the same linear density?

(7) Estimate the difference in bending stiffness between Lincoln and Merino wool as best you can.

(8) A PET fiber has the cross-sectional area shown:

State as many techniques as you are able that may be used to determine the fiber denier.

(9) A round PP fiber has a density of 0.93 g/cm^3 and a denier of 0.30. Calculate the fiber diameter in μm and cross-sectional area in $(\mu m)^2$.

(10) Derive an expression for the specific external surface area as a function of fiber inner and outer radius for round, hollow fibers.

(11) A wool fiber is properly positioned in a vibrascope and loaded with a 200 mg weight at a test length of 2 cm. The fundamental frequency of vibration is 1355 Hz (cycles/sec). Calculate the denier and diameter of the fiber, assuming wool is essentially round with a density of 1.30 g/cm^3.

(12) Suppose you have one cotton plant loaded with ripe cotton balls. A textile scientist offers you $100 if you bring him only the immature or dead cotton. Do you accept his offer?

(13) Derive an expression for the specific surface area of a rectangular fiber with $w = h \times 5h$.

(14) A vibrascope experiment is conducted on a cotton fiber using a hung mass of 100 mg and a fiber length of 2 cm. When the experiment is repeated using 400 mg and a length of 1 cm, how does the fundamental frequency change?

(15) For a new vegetable fiber with a rectangular cross-section $h \times 3h$ and density 1.30 g/cm^3, calculate the surface area-to-volume ratio compared with that of a round fiber of equal cross-sectional area.

(16) What experimental technique might you select to provide data on the average denier of
  (a) 10 fibers each 3 cm long?
  (b) 1000 fibers each 3 cm long?
  (c) 1000 hollow fibers each 3 to 5 mm long?

(17) For the cotton fiber shown magnified 1200X, determine

(a) specific surface area.

(b) denier.

(c) maturity.

Assume a typical density for cotton.

(18) Show that the surface area-to-volume ratio varies with $1/r$ for round fibers; that is, the ratio blows up at $r = 0$.

(19) A sample of polyethylene is melt spun to give a trilobal fiber. Using optical microscopy, a coworker determines that the cross-section of the fiber can be accurately represented as an equilateral triangle with each side length = 24.3 micrometers. The fiber is highly uniform.

(a) Calculate the specific surface area of the fiber in units of reciprocal μm.

The trilobal fiber is placed in a vibrascope. The test length is 1 inch, a 76 mg weight is suspended from the fiber, and the fundamental frequency is found to be 1067 cycles/sec.

(b) Determine the fiber denier. Determine the density of the fiber in g/cm^3.

6

Environmental Effects: Solvents, Moisture, and Radiation

Textile and industrial fibers respond to the environment according to their chemical composition and microstructure. Since fibers are typically semicrystalline, they are generally more stable to both chemical attack and temperature than are wholly amorphous linear polymers. Polymer fibers are very good; indeed, they are the material of choice in some applications, yet they perform poorly in other environments. For example, polyolefins are virtually inert to inorganic acids, making polyolefin fibers useful as battery separators; however, polyolefins, like most other polymers, do not resist degradation in ultraviolet light. Consequently, polyolefin tent fabric is stabilized so that it can tolerate long exposures to sunlight. We will probe the cause of poor ultraviolet stability in polymers and discuss ways to stabilize fibers.

One of our tasks in this chapter is to investigate the effect of various liquids and vapors on the structure and properties of fibers. When a fluid has a chemical structure similar to that of a solid polymer, there is a high likelihood that the liquid will diffuse into the polymer at a rate that is determined by both chemical and physical structure. If the chemical structure of the liquid or vapor is dissimilar to that of the polymer, then the polymer will generally not absorb the fluid or vapor. This rule of thumb is "like dissolves like." It applies to all materials. Consider, for example, oil and water. Water has the structure shown on the next page. Oxygen has two lone pairs of electrons, and it is electronegative. Hence, water is a polar molecule.

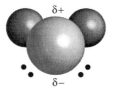

It binds to and forms solutions with other polar materials. Oil, on the other hand, is a hydrocarbon consisting essentially of short chains of polyethylene, CH_3—CH_2—$\cdots$—CH_3. There is no dipole moment associated with saturated hydrocarbons, so the molecules are not polar. The dissimilarity in structure suggests that water and oil do not mix, a rather well-known fact.

6.1 Solubility Parameters

We need to be more quantitative than simply saying like dissolves like. The term we use is *solubility parameter*. It is a measure of the inherent capability for secondary bonding to a molecule, and therefore relates to the heat of melting or vaporization. Solubility parameters can be calculated and assigned to various organic functionalities. Materials—polymers and liquids or vapors— with similar solubility parameters have an excellent chance of interaction. Materials with highly dissimilar solubility parameters are typically nonreactive. Some materials, such as PET, have more than one solubility parameter. PET has a solubility parameter associated with the terephthalic acid residue and another with the ethylene glycol residue. The former is more polar than the latter. Fluids that match either solubility parameter may interact with PET; mixed solvents may also dissolve PET. In addition, polymers with similar structure or solubility parameters as other polymers have the best chance of forming single-phase solid solutions or polymer alloys. A chart of solubility parameters may be found in the *Polymer Handbook*, listed in the reference section. A partial listing of some fluids and polymers is provided in Table 6.1.

The solubility parameter concept alone does not treat all the factors that are associated with solubility. Hence, it does not give the correct answer all the time. For example, it does not differentiate polar from nonpolar interactions. Nonpolar solvents will not interact with polar polymers, and polar solvents will not interact with nonpolar solvents, regardless of the closeness of the solubility parameters. Most scientists, however, use solubility parameters as a starting point when they are required to identify potential solvents or nonsolvents for a new or unfamiliar material.

Solvation does not involve degradation of the polymer molecular weight or alteration of any primary bonds. Cellulose is a very difficult material to dissolve because of the concerted effect of the hydroxyl groups. Many "solvents" for cellulose actually cause molecular weight degradation. Similarly, wool, a polyamide, is not just dissolved by alkali, such as bleach; rather, it is degraded.

It is instructive to note the similarities and differences between melting and dissolution. Melting is the breaking up of secondary bonds in crystalline

TABLE 6.1 Solubility Parameters of Various Polymers and Fluids

Material	Solubility Parameter
	$(J/m^3)^{1/2} \times 10^{-3}$
Fluids	
propane	13.1
water	47.9
acetone	20.3
dimethyl formamide	24.8
ammonia	33.4
toluene	18.2
dioctylphthalate	16.2
formic acid	24.8
nitrobenzene	20.5
Polymers	
PET	21.9
nylon	~ 25
polystyrene	18.6
acrylic	25.6
cellulose	32.0
cellulose acetate	27.8
polytetrafluoroethylene	12.7
polybutadiene	17.0
PE	16.2
PP	19.0

Source: J. Brandrup and E. H. Immergut, eds., *Polymer Handbook*, 3rd ed., New York: Wiley-Interscience, 1989.

regions. The breakup is facilitated by the addition of thermal energy. Similarly, dissolution is replacement of the secondary bonds among polymer chains with secondary bonds between polymer and solvent. Dissolution involves immersing a polymer in a sea of solvent. Melting is essentially solvation into molten molecules of itself. Neither involves degradation of primary bonds.

In courses on polymers, you will learn about good solvents and poor solvents. A good solvent is one that readily and easily dissolves the polymer. A poor solvent is one that is on the brink of not being able to dissolve the polymer. Flory defined a polymer–solvent interaction parameter, χ, which characterizes the interaction energy per mole of solvent for each solvent–polymer pair (Flory, 1953). χ has been empirically related to solubility parameter, but it is best to determine χ experimentally.

Solvents can dissolve only linear polymers. When a polymer is cross-linked, a fluid can at best swell the polymer. Covalent bonds hold the polymer molecules together, and solvents disrupt only secondary bonds. The swelling is caused by the solvent wanting to associate with the polymer segments, form secondary bonds, and in so doing, separate the molecules. On the other hand, the cross-links oppose the separation. Thus, Flory's equilibrium swelling equation has a mixing term on the left and an elastic term on the right:

$$\ln(1 - v_p) + v_p + \chi v_p^2 = -NV_s(v_p^{1/3} - v_p/2)$$

where

v_p = volume fraction polymer in the mixture,

N = number of polymer chains per unit volume,

V_s = molar volume of the solvent, and

χ = polymer–solvent interaction parameter, which can be mathematically related to the solubility parameter.

Using this equation, the molecular weight between crosslinks, N, and hence the crosslink density, may be determined. At this point it is not critical that you be able to use the swelling equation, but you should appreciate that cross-linking prevents solubilization of a polymer.

6.2 The Interaction of Fibers with Moisture

Water and water vapor are highly polar materials. We therefore anticipate that moisture will diffuse into polar polymers, such as cotton, nylon, and wool. Our expectation is completely borne out by observation and measurement. All natural fibers—both animal and vegetable—are highly polar and absorb significant quantities of water. On the other hand, the polyolefins and other synthetic nonpolar polymers do not absorb water. When water is absorbed into a fiber, a number of phenomena occur. The formation of new, strong secondary bonds facilitates the liberation of heat; the water diffuses from the surface to the center of the fiber following specific kinetic laws; and the physical properties of the fiber change with the absorption of moisture. Consider what processes occur when you take a dry fiber and expose it to an environment containing a high amount of water vapor. The water vapor adsorbs and condenses onto the surface of the fiber, causing a change in the secondary bond structure of the polymer, heating of the surface material, and a change in various properties. The water diffuses through the swollen polymer and local heating continues as the front diffuses to the center of the fiber. This swelling process ends when the fiber is completely saturated with the equilibrium amount of water. Indeed, absorption is a complex event.

We will divide the absorption process into four parts and treat each one as separate in this section. Such division is clearly an approximation, as the processes interact. We first introduce some definitions that facilitate discussion of moisture effects. Then, in the following three sections, we discuss various aspects of moisture absorption—thermal effects, rate of absorption, and property changes associated with moisture absorption. Keep in mind that there is nothing really special about the interaction of fibers with water. What you learn in this chapter applies equally to all fibers and all fluids or vapors. We simply discuss water as an important example, since water is everywhere in the atmosphere, the moisture content of the air changes all the time, and many fibers absorb water.

6.2.1 Definitions: Moisture Regain and Content

In this and subsequent sections we are concerned with moisture absorption, which is the bulk uptake of water into a material. We will not dwell on adsorption, which is the formation of a thin layer of moisture on the surface of a material. The first concept that needs to be understood is that of relative humidity. Let us begin by discussing vapor pressure. The amount of moisture in the air is measured as vapor pressure. The maximum vapor pressure of moisture in air increases with temperature. Hence, warm air can hold considerably more moisture than can cold air. The dew point is the temperature at which the moisture in the air will begin to precipitate, or rain, because the drop in temperature causes a decrease in air's ability to hold moisture. Humidity is the mass of moisture in the air per unit volume of air. *Relative humidity* is a more useful term than *humidity* for us because it accounts for the temperature dependence of the moisture-carrying capacity of air. Relative humidity is the amount of moisture in the air compared with the amount of moisture the air can hold at that temperature. RH can never exceed 100 percent, because at that moisture content, any added moisture vapor will precipitate as liquid (rain).

As moisture is added to a dry fiber, the mass of the fiber increases. The mass of the absorbed water divided by the mass of the dry fiber is called moisture regain, R:

$$R = \text{mass moisture absorbed/mass dry fiber}$$

Another way to characterize the amount of moisture in a fiber is with moisture content, M. It is defined as the mass of moisture absorbed normalized to the mass of the wet fiber:

M = mass moisture absorbed/mass of wet fiber

    = mass moisture absorbed/(mass moisture absorbed + mass dry fiber)

M and R are not independent. Consider, for example, a dry fiber that absorbs its weight in moisture. $R = 1.0$ and $M = 0.5$. The relationship may be generalized:

M = mass moisture absorbed/(mass moisture absorbed + mass dry fiber)

    $= \text{MM}/(\text{MM} + \text{MF}) = 1/(\text{MM}/\text{MM} + \text{MF}/\text{MM})$

    $= 1/(1 + 1/R) = R/(R + 1)$

Equilibrium Moisture Regain and Content

When a dry fiber is subjected to an environment containing moisture, the fiber absorbs moisture at a rate that depends on a number of physical factors. The

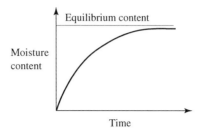

FIGURE 6.1   Achievement of equilibrium moisture regain.

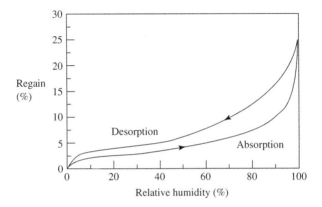

FIGURE 6.2   Effect of relative humidity on equilibrium regain of cotton. (A. R. Urquhart and N. Eckersall, *Journal of the Textile Institute* 15 (1930), T499.)

initial stage of moisture uptake is called the transient stage. At some time the moisture content begins to reach a constant value, called the equilibrium value, as shown in Figure 6.1. Note that the rate of change decreases with time until at equilibrium there is no change in regain with time.

Now that we understand the concept of equilibrium regain, let us discuss the effect of relative humidity on equilibrium regain. We anticipate that R will increase with RH, since concentration is the driving force. The change in equilibrium regain for cotton with RH is shown in Figure 6.2. We will use the equilibrium moisture content or regain most of the time in this text to characterize fibers. To facilitate comparison of various materials, we select 20 °C and 65 percent RH as standard conditions. Table 6.2 shows the regain of important textile fibers at standard conditions.

Fibers with polar groups show high regain, and nonpolar fibers show no regain. A question that many students ask when seeing these data is why the regain for wool is about double that of cotton. Recalling the structures, you note that both contain a high level of polar groups. The key to the answer lies not with capacity for water, but with accessibility of water. Cotton is very highly crystalline, i.e., on the order of 100 percent distorted and defective crys-

TABLE 6.2   Moisture Regain of Textile Fibers at Standard Conditions

Fiber	Absorption Regain (%)	Hysteresis (%)
Natural fibers		
Cotton	7.5	0.9
Mercerized cotton, jute	12	1.5
Flax	10	—
Viscose rayon	13	1.8
Secondary acetate	6–7	2.6
Silk	10	1.2
Wool	14–18	2.0
Synthetic fibers		
Nylon 6 or 66	4.1	0.25
PET	0.4	—
Acrylic	1–2	—
Polyvinyl alcohol	4.5–5	—
Polyolefin	0	0
Oxide Glass	0	0

Sources: J. E. Ford, *Fiber Data Summaries*, Manchester, England: Shirley Institute, 1966; J. Brandrup and E. H. Immergut, eds., *Polymer Handbook*, 3rd ed., New York: Wiley-Interscience, 1989; J. B. Speakman and E. Scott, *Journal of the Textile Institute* 27 (1936), T186.

tals. Wool, on the other hand, is roughly 40 percent crystalline. The interaction with moisture is not sufficiently great that it can completely disrupt the crystalline regions in cotton.

Water is indeed not an excellent solvent for cellulose or silk. If water were a particularly good solvent, it might completely dissolve silk or cotton, which lack crosslinks. A good solvent for cellulose has been the object of extensive research for many decades. Cellulose is usually modified in order to achieve dissolution for the making of rayon fibers. The chemicals used in the viscose rayon process are neither polymer-friendly nor environmentally favorable. New solutions and processes are required to facilitate the formation of continuous fibers from cellulose. Courtaulds, in fact, has recently opened a rayon plant that uses n-methyl morpholine oxide or something similar (N. Franks, US Patents 4,196,282 (1980); 4,145,532 (1979); 4,136,255 (1979)).

Industrial fibers usually have low regain, partly because of their high orientation and crystallinity; however, in many applications it is a requirement that fiber properties not change with atmospheric conditions. Fiber chemistry is selected accordingly.

Hysteresis in Moisture Uptake

The equilibrium value of moisture content or regain has a unique feature. It depends on the direction of approach, as shown in Figure 6.3. When the equilibrium value is approached from the high-humidity side, or during desorption,

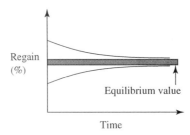

FIGURE 6.3 Effect of direction of approach of moisture regain.

the value of regain or content is greater than when the equilibrium value is approached from lower RH or during absorption. The right-hand column in Table 6.2 shows the hysteresis, or difference in absorption and desorption equilibrium values at standard conditions for various fibers. Those of you familiar with equilibrium phenomena will realize that true equilibrium is *not* dependent upon the direction of approach. Hence, the phenomenon of hysteresis in fibers suggests that there is something other than thermodynamic processes occurring during moisture sorption. In fact, this assumption is correct. Moisture absorption causes structural changes in the fiber. The structural changes are responsible for the hysteresis.

Hysteresis is a general term alluding to the fact that a physical property is path-dependent. Hysteresis generally occurs when losses of some sort occur. Graphically, hysteresis loss appears as you see in the absorption–desorption curves. The curve of increasing *y* versus *x* differs slightly from that of decreasing *y* versus *x*. We will see hysteresis in elastic response in Chapter 7 as well as in other places in this text.

6.2.2 Heat of Moisture Absorption

Absorption of water may occur from the liquid, the vapor, or both. When water is absorbed from the vapor, it first effectively condenses onto the surface of the fiber. The latent heat of condensation—the heat released on changing phase from the vapor to the liquid state—is large, about 2450 J/g. This effect can dwarf the other thermal effects we discuss subsequently. When the relative humidity reaches or exceeds about 99 percent, then not all the condensed water is absorbed by the fibers. Some condensate remains on and between the fibers as capillary water.

Molecules form secondary bonds because doing so reduces the total energy of the system. The molecules in cotton or wool have a difficult time arranging themselves in a conformation that maximizes secondary interactions. Crystals allow for optimal secondary bond interaction, but that in the noncrystalline areas has room for improvement. When a liquid or vapor capable of satisfying secondary bonding is introduced into the fiber, the molecules in the noncrystalline regions make minor conformational adjustments to maximize secondary

TABLE 6.3 Heat of Wetting of Various Fibers from Zero Regain

Fiber	Heat of Wetting (J/g)
Natural fibers	
Cotton[a]	46
Mercerized cotton[a]	73
Viscose rayon[b]	106
Acetate[b]	34
Wool[c]	113
Silk[d]	69
Flax[b]	55
Synthetic Fibers	
Nylon 6 or 66[c]	31
PET[e]	5
Acrylic[c]	7

Sources: [a]J. Brandrup and E. H. Immergut, eds., *Polymer Handbook*, 3rd ed., New York: Wiley-Interscience, 1989; [b]J. C. Guthrie, *Journal of the Textile Institute* 40 (1949), T489; [c]N. F. H. Bright, T. Carson, and G. M. Duff, *Journal of the Textile Institute* 44 (1953), T587; [d]J. J. Hedges, *Transactions of the Faraday Society* 22 (1926), 178; [e]W. E. Morton and J. W. S. Hearle, *Physical Properties of Textile Fibers*, 3rd ed., Manchester, England: The Textile Institute, 1993.

interactions. By improving the extent of secondary bonding, the energy of the system is reduced and the excess energy is liberated in the form of heat. This is the heat of absorption. You may envision the heat of absorption as the heat released when water reacts with the polymer molecules, the reaction being the formation of hydrogen bonds between polymer and water. A similar effect is observed when any interacting fluid comes in contact with a polymer. It is called the heat of solvation.

Integral Heat of Absorption

Two expressions are commonly used to quantify the amount of heat released (exothermic) upon absorption or required (endothermic) for desorption. The first is the integral heat of absorption, also called the heat of wetting. It is the total amount of heat given off when an infinite amount of water is added to fiber whose dry mass is 1 g. The sample may be dry to begin with or it may be partially hydrated. The units of heat of absorption are J/g of fiber. Typical values for textile fibers are given in Table 6.3. The data range from a low of 5 (a value of 0 is possible) to a high of 113 J/g for wool. The nil value is for fibers that do not absorb water. The high values are for the fibers that absorb the most water. When the heat of wetting is divided by the regain, the results show that nothing unusual happens when wool absorbs water. The high value for heat of absorption is simply due to the fact that wool is characterized by high regain. Sheep and garments made from wool are quite warm in a cold rain, since absorption of the rain produces heat. Conversely, a cooling effect is observed when the bound moisture desorbs from the material.

TABLE 6.4   Differential Heats of Sorption of Various Fibers (kJ/g)

Fiber	0	15	30	45	60	75
Natural						
Cotton[a]		1.24	0.50	0.39	0.32	0.29
Mercerized cotton[b]	1.17	0.61	0.44	0.33	0.23	
Viscose rayon[a]	1.17	0.55	0.46	0.39	0.32	0.24
Acetate[a]	1.24	0.56	0.38	0.31	0.24	
Wool[c]		1.34	0.75	0.55	0.42	
Synthetic						
Nylon 6 or 66[d]		1.05	0.75	0.55	0.42	

Sources: [a] J. C. Guthrie, *Journal of the Textile Institute* 40 (1949), T489; [b] W. H. Rees, *Journal of the Textile Institute* 39 (1948), T351; [c] N. F. H. Bright, T. Carson, and G. M. Duff, *Journal of the Textile Institute* 44 (1953), T587; [d] J. B. Speakman and A. K. Saville, *Journal of the Textile Institute* 37 (1946), P271.

Differential Heat of Sorption

The differential heat of sorption is the amount of heat given off when 1 g of water is added to an infinite amount of fiber of given regain. The units of differential heat of sorption are J (or kJ)/g of water. Typical values of differential heat of sorption are provided in Table 6.4. The values in the table are the largest for fibers with low initial moisture content. This is because the first water binds the most tightly, releasing a large amount of heat. (In fact, the differential heat of sorption at 0 percent RH is about the same as that for the heat of hydration of hydroxyl ions, suggesting the similarity between this experiment and the addition of water to an acid.) The last water to bind is bound loosely. Any additional water fills the pores or capillaries, and it is unbound, or so-called free water. Water begins to collect in the spaces between fibers, the capillaries, when the relative humidity reaches about 99 percent.

6.2.3 Rate of Moisture Absorption and Desorption

So far we have focused on equilibrium moisture absorption. Fiber scientists must have knowledge of the rate of moisture absorption and desorption. Suppose you are purchasing a large quantity of wool. The wool has been stored at 30 °C and 90 percent RH for several months. It has been only recently moved to a storage facility at 20 °C and 65 percent RH. It is important that you know just how much water and wool you are purchasing, so that you do not pay too much.

Diffusion of Moisture

The rate that moisture diffuses into or out of a fiber or any solid is governed by Fick's First Law, which states that the mass of moisture passing through a cross-section, called the flux, J, is proportional to the concentration gradient

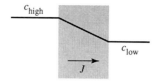

FIGURE 6.4 Diffusion through a membrane.

of moisture, dc/dx. The greater the rate of change of moisture content across a cross-section, the greater the flux:

$$J \propto dc/dx \Rightarrow J = -D(dc/dx)$$

The proportionality constant is D, the diffusion coefficient. D has units of cm^2/sec; D may be determined experimentally or looked up in tables. The negative sign in the equation indicates that moisture moves from a region of high concentration to one of low concentration, as shown in Figure 6.4. Fick's First Law has limited use, as we shall see; however, one practical problem that can be solved using Fick's First Law is the membrane problem. A membrane separates two flowing streams of gasses that have different concentrations of an impurity. Fick's Law can be used to determine the steady-state flux of impurity through the membrane, assuming D is a known quantity:

$$J = -D(\Delta c/\Delta x)$$

where

Δc = the concentration difference, $c_H - c_L$,

Δx = the thickness of the membrane,

D = diffusion coefficient of the impurity through the membrane, and

J = flux of impurity through the membrane per unit area of membrane.

Fick's First Law is useful here because the concentration gradient is constant, and at steady state D is constant. Most problems in diffusion, such as our fiber problem, have a region where c changes with time. For example, a fiber experiences an increase in moisture content during absorption. In these cases, we cannot use Fick's First Law, but another expression is needed that describes the flux that corresponds to the experimental conditions—Fick's Second Law.

Consider the diffusion of an impurity through a solid bar, as sketched in Figure 6.5. The flux through the bar of unit cross-sectional area at x coordinate 1 can be expressed using a Taylor series expansion of two terms—J_2 minus a differential amount:

$$J_1 = J_2 - \Delta x(\partial J/\partial x) \Rightarrow J_1 - J_2 = \Delta x(\partial J/\partial x)$$

Since $J_1 \neq J_2$, the concentration of species, c, in the volume Δx changes with time:

$$J_1 - J_2 = \Delta x(\partial c/\partial t)$$

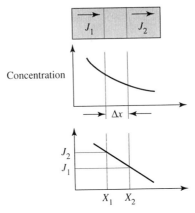

FIGURE 6.5   Diffusion of impurity through a bar.

Setting the two expressions for ΔJ equal is a statement of the mass conservation law,

$$\partial c/\partial t = \partial J/\partial x$$

Substituting the expression for J from Fick's First Law,

$$\partial c/\partial t = \partial(D\partial c/\partial x)/\partial x$$

we get Fick's Second Law. Sometimes you will see Fick's Second Law in a simpler form, which may be used when the diffusion coefficient is not a function of x or, therefore, c:

$$\partial c/\partial t = D(\partial^2 c/\partial x^2)$$

Fick's Second Law completely describes diffusion of species. We use it to determine concentration as a function of time. It is a second-order partial differential equation. The specific solution to a real one-dimensional diffusion problem described by Fick's Second Law requires knowledge of one initial and two boundary conditions. We will be concerned with only one relatively simple class of approximate solutions. Our problem is to determine about how long it will take a fiber to absorb or desorb moisture. We address this problem in the next section.

Before proceeding, let us extend our knowledge simply by making an analogy. A process similar to moisture or mass transfer is heat transfer. Like mass, heat diffuses from one position to another down a gradient, a temperature gradient in this case. The mathematics of thermal diffusion are identical to those of mass diffusion. Only the variables differ. Consider, for example, the parallel between Fick's First Law of Diffusion and Fourier's Law of Conduction:

Fick's Law: $J = -D(\partial c/\partial x)$

Fourier's Law: $q = -k(\partial T/\partial x)$

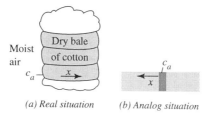

(a) Real situation   (b) Analog situation

FIGURE 6.6 Schematic of conditioning process.

where

q = heat flux per unit area,

T = temperature, and

k = thermal conductivity, a material property.

In Fourier's Law, k, a material property like D, is the proportionality constant between (heat) flux and (temperature) gradient. There is also an equation directly analogous to Fick's Second Law, called the Heat Equation:

Fick's Second Law: $\partial c / \partial t = D(\partial^2 c / \partial x^2)$

Heat Equation: $\partial T / \partial t = \alpha(\partial^2 T / \partial x^2)$

where $\alpha = k/\rho C_p$, the thermal diffusivity. (C_p is heat capacity and ρ is density.) The point of all this is that heat flow and moisture transport are analogous. The concepts are analogous, the mathematics are identical, and the solutions to the mathematics are the same. Only the numbers differ. Such analogies provide an easy way to learn.

Fiber Conditioning for Moisture Uptake

Recall the problem discussed earlier in this chapter, that of changing the environment in which a fiber sample is stored. How long does the sample require to reach moisture or thermal equilibrium in the new environment? Fiber scientists need to be able to calculate the approximate time it requires to reach equilibrium. We do the analysis using Fick's Second Law. We are not seeking a highly accurate solution to the problem; rather, an order of magnitude estimate of the time involved is sufficient for our needs. We proceed by noting that the problem consists of changing from one uniform moisture content to another, as illustrated in Figure 6.6a. At this time we are not especially concerned with the entire curve representing the time dependence of moisture content as a function of position in the fiber. Rather, we are interested in calculating the time required for a sample to reach equilibrium when exposed to a new environment.

The problem will be treated by considering the analogous one-dimensional problem, illustrated in Figure 6.6b. A bar of material is cut in half. Another material is coated onto one end of the bar, and the two ends are reassem-

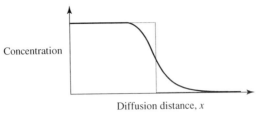

FIGURE 6.7 Graphical solution to the diffusion problem showing the concept of penetration distance.

bled. The bar could be copper and the plated metal might be silver. Alternatively, the bar could be nylon and the coating could be water. The solution to this problem, diffusion into a semi-infinite medium, has been worked out (Incropera and De Witt, p. 909) and is of the form

$$(c - c_{surface})/(c_{initial} - c_{surface}) = erf[(x/(4Dt)^{1/2}]$$

where

erf = the Gaussian Error Function, which is tabulated in many texts,

$c_{surface}$ = the concentration of the impurity on the surface, a constant,

$c_{initial}$ = the concentration of the impurity intially in the bar, and

c = time-dependent concentration of impurity at position x.

A graphical representation of the solution to the preceding equation is shown in Figure 6.7. We see that the concentration is low until the "front" passes through, whereupon the concentration rapidly achieves a high value. What we effectively do in these sorts of problems is simply determine the time required for the "front" to pass through. This is equivalent to approximating the curve in Figure 6.7 as the dotted line. When $x/(4Dt)^{1/2} \ll 1$, the front has not arrived. When $x/(4Dt)^{1/2} \gg 1$, the front has passed. Therefore, when the argument of the error function, $x/(4Dt)^{1/2} = 1$, we say the front has arrived. To determine the approximate time for a fiber to reach equilibrium, we simply calculate the time required to satisfy

$$t \approx x^2/4D$$

We need only the diffusion coefficient, and the distance moisture needs to diffuse in order to penetrate the most inaccessible part of the sample. An example of how to use this simplified form of the diffusion equation is provided in the following section. You will often see the expression

$$t \approx x^2/D \text{ (mass)} \quad \text{or} \quad t \approx x^2/k \text{ (heat)}$$

with the factor 2 or 4 omitted. Since the analysis is used for order-of-magnitude estimates only, the constants are considered negligible.

Thus, the calculation to determine the time required to reach equilibrium is the time required for penetration to the most distant point. It depends on the thermal or mass diffusion coefficient and the sample geometry. A typical value for the diffusion of moisture through a natural fiber is roughly 10^{-7} cm^2/sec. A reasonable value for fiber radius is 10^{-3} cm. Substitution and calculation indicate that about 5 seconds are required for a single fiber to reach moisture equilibrium. (A similar calculation shows that it requires $\ll 1$ second for a fiber initially at 20 °C to thermally equilibrate when immersed in an oven at 200 °C.) Experimental measurements show that a single fiber requires about 1 to 5 minutes to reach moisture equilibrium. Our calculation was within an order of magnitude, as designed. The experimental value exceeds the calculated one largely because D *does* change substantially with moisture content and other thermal, structural, and chemical factors, which were neglected in the simple analyses, are involved.

The same sort of calculation may be conducted on an aggregate of fibers. Consider, for example, a tightly packed bale of cotton fiber or a tightly twisted cord of nylon rope. In these cases the surrounding air cannot effectively penetrate the bale or cord. An upper-bound calculation may be made, assuming that the only way moisture can get into or out of the aggregate is by diffusion. The diffusion path is much longer than that into the single fiber. Since $t \propto x^2$, it takes a long time to reach equilibrium. Consider a cubic bale that is 1 meter on a side. D is the same as just used, 10^{-7} cm^2/sec. Substitution gives a time of 8 centuries. A lower-bound calculation may be conducted assuming the moisture diffuses 0.5 m through still air and then through the fibers. Using a diffusion coefficient of 0.25 cm^2/sec for moisture in still air, the time to penetrate the bale is 3 hours. Experimental results show that the actual time to condition a bale of this size is roughly 3 days, which is much closer to the lower limit than the upper limit. The diffusion path in the air is certainly not a straight line, but rather, a tortuous path among the fibers.

The calculations just performed are based on a number of simplifications and approximations. This is the reason the estimates are order-of-magnitude only. Recall that the uptake of moisture involves not only diffusion, but structural rearrangements, liberation of heat, and fiber swelling; in addition, D is a function of moisture concentration. For example, isothermal conditions were assumed, yet it is not unusual for the temperature of cotton to rise 10 °C or more with a rapid increase in RH of about 50 percent. It is in fact remarkable that the calculations work out so well in light of the crudeness of a number of the assumptions.

6.2.4 Physical Property Changes with Moisture Uptake

When fibers take up moisture or any other liquid, virtually all of the properties of the fiber change. We discuss many of the property changes in the course of this text. In Chapter 14 we discuss the changes in electrical conductivity that accompany increases in moisture content. The concept of diffusion facilitates

TABLE 6.5   Swelling of Fibers in Water

Fiber	Cross-Sectional Area (%)	Axial (%)	Volume (%)
Natural			
Cotton	42	0.15	
Mercerized cotton	46	0.1	
Viscose rayon	50–115	3.7–4.8	109–127
Jute	40		
Silk	19	1.5	31
Wool	25	1.4	37
Synthetic			
Nylon 6 or 66	1.6–3.2	2.7–6.9	8–11

Source: J. M. Preston and M. V. Nimkar, *Journal of the Textile Institute* 40 (1949), P674.

an understanding of the reason absorbed moisture improves the electrical conductivity by several orders of magnitude in many textile fibers, which are normally considered to be insulators. In Chapter 7 we learn about the basic mechanical properties of textile and industrial fibers. As we have described, moisture, like many other organic fluids, absorbs into many textile fibers. The process of fluid absorption is often referred to as plasticization, especially when mechanical properties are being discussed. The plasticizer softens the polymer and changes a number of the mechanical properties. For example, vinyl seat covers contain about 40 percent of an organic fluid called dioctyl phthalate, DOP, which softens the vinyl. Without DOP, vinyl seat covers would be stiff and brittle, much as they are in old cars after the DOP has diffused out, assisted by prolonged exposure to hot–cold cycles and sunlight. The following expression has been found useful in calculating the effect of plasticizer content on the glass transition temperature of noncrystalline polymers:

$$T_g^{-1}(\text{solution}) = w_1 T_g^{-1}(\text{polymer}) + w_2 T_g^{-1}(\text{plasticizer})$$

where the w's are weight fractions of polymer (1) and plasticizer (2). Physically, the fluid surrounds the polymer molecules, enhancing the mobility of the mers and reducing the glass or softening temperature.

Volume Swelling

The introduction of moisture or any solvent into fibers may cause increases in fiber volume. When the moisture simply fills pores, as when a paper towel is used to quickly wipe a spill, little volume change occurs; however, when fibers absorb moisture, the moisture diffuses into each fiber, causing the molecules to separate and the fiber volume to increase. The results of moisture swelling measurements on a variety of fibers are shown in Table 6.5. The data show that volume increases are substantial and that the volume increase is chiefly

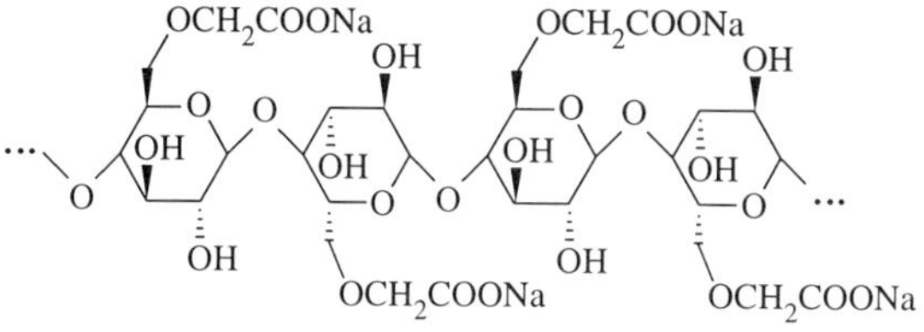

FIGURE 6.8 Structure of carboxy methyl cellulose.

due to increases in fiber cross-sectional area. Fiber length changes are small. The exceptions to this behavior are rayon and nylon. It has been suggested that in these cases a skin restrains the cross-section from increasing.

Solvent swelling is a powerful technique in polymer analysis. When polymer and fluid have similar solubility parameters, the polymer will likely take up solvent. When the polymer is noncrystalline, it may completely dissolve rather quickly; however, when the polymer is crosslinked, like wool, it will only swell. The extent of swelling is directly related to the extent of crosslinking. When the polymer is semicrystalline, the noncrystalline areas swell and sometimes the crystalline material will then slowly dissolve.

A great triumph in fiber science was the recent development of superabsorbent fibers. Imagine what would occur if a polymer with a structure similar to cellulose but with no crystallinity and fewer hydrogen bonds were exposed to water. Consider the modified cellulose, carboxy methyl cellulose, CMC, shown in Figure 6.8. The bulky side groups reduce the capacity for hydrogen bonding and also preclude the ability to crystallize. With no crystallites to limit swelling and fewer hydrogen bonds present, moisture absorption is expected to be much higher. Water will in fact dissolve CMC, so the polymer is lightly crosslinked. The lightly crosslinked particle or fiber of CMC is now capable of absorbing more than 100 times its weight in water, forming a hydrogel. Other superabsorbent materials, both modified natural and synthetic, are also available. These materials have partially replaced conventional absorbents in diapers, such as wood pulp fibers, or fluff, and have revolutionized the disposable diaper industry.

Density Changes

We now realize that the introduction of moisture causes both the mass and volume of a fiber to increase. How then does the density, or the mass per unit volume, change with moisture uptake? The answer lies in the details associated with swelling. At 65 percent RH, cotton has a density of 1.55 g/cm^3, whereas dry it has a density of 1.52 g/cm^3, suggesting the mass increase outweighs the volume increase. Similarly, wool has a density of 1.31 g/cm^3 at 65 percent RH yet only 1.30 g/cm^3 dry. Viscose rayon, on the other hand, is characterized by a lower density at 65 percent RH than dry, i.e., 1.49 versus 1.52 g/cm^3. Most other hydrophilic fibers, as do the hydrophobic fibers, have densities that are essentially invariant to atmospheric moisture level changes.

6.3 The Effects of Radiation on Fiber Structure and Properties

Polymers are notoriously unstable to ionizing or even ultraviolet radiation. Incident UV light has sufficient energy that impurities present catalyze decomposition of the polymer. The radicals produced photolytically degrade the molecular weight of the polymer, as shown in Figure 6.9. The active chain ends produced by photolytic degradation may react with the polymer further, producing crosslinks that embrittle the polymer. As the changes in molecular weight continue, the mechanical, optical, and other properties of the polymer are seriously degraded. Fibers that will be subjected to prolonged exposure to ultraviolet or more energetic radiation need to be protected. When, for example, a high-density polyethylene milk jug is left in the Georgia sun for one year, the once-tough jug becomes quite brittle. The strength of the polymer has been seriously compromised. Similarly, when wool is subjected to prolonged sunlight, the weakest bonds, the disulfide linkages, —S—S—, are attacked first. Human skin is a natural polymer. It too degrades with prolonged exposure to sunlight. We call it tanning or burning.

Fibers that have been exposed to UV radiation for prolonged periods of time show visual effects of the degradation. Scanning electron micrographs typically show pits or grooves, and eventually, with significant loss of material, an interconnected pore structure develops. The presence of moisture assists the degradation of hydrophilic fibers. There is compelling evidence that TiO_2, titanium dioxide, which is often added to whiten or opacify fibers, is a UV sensitizer. Degradation begins or occurs at a faster rate in the immediate vicinity of TiO_2 particles than in regions of pure polymer.

One means to afford protection to polymers from electromagnetic radiation is to paint, metallize, or otherwise coat the surface of the polymer to prevent the UV light from penetrating the polymer. For various reasons, this method is not particularly attractive to fiber scientists. Another way to protect fibers is to add stabilizers. The stabilizers come as packages designed for use with a specific polymer. A stabilizer package is simply a combination of chemicals that postpones thermal or ultraviolet degradation. The additives contain chemicals that harmlessly absorb the destructive radiation. Sometimes the additives absorb the harmful UV radiation and reradiate in the blue region. These additives are called bluing agents or optical brighteners, and make a fiber appear "whiter." The packages are added to synthetic polymers prior to fiber

(a) 3 molecules   (b) Scission→ 8 molecules   (c) Crosslinking→ 1 molecule

FIGURE 6.9   Schematic of photodegradation in polymers.

formation. The stabilizers may also prevent thermal breakdown during melt extrusion or fiber processing. Smartly formulated stabilizer packages are quite effective, allowing such UV-sensitive polymers as PE, PP, and PET to be used as tent fabric, boat covers, and in playground equipment. Similarly, natural fibers can be imbued with various stabilizers to enhance their useful lifetime in the environment.

6.4 Summary

Polymers are generally soluble in fluids that have a structure similar to that of the polymer. Hence, solubility can be predicted or rationalized on the basis of similar solubility parameters. Water is a special material, since it is present in the air around us as vapor or rain, and it is the fluid with which we clean our laundry. While only a few polymers are soluble in water, many polymers absorb moisture into their internal structure. These polymers can be identified by their chemical structure. Important fibers that absorb water are the natural fibers, plus other nylons. All are capable of hydrogen bonding. The amount of moisture uptake depends on the number and strength of hydrogen bonds that can be formed with water and on the fiber crystallinity. Secondary bonding in crystalline polymer may be strong and difficult to disrupt. Hence, penetration into and diffusion through crystalline regions is much slower than through noncrystalline regions.

Diffusion of water is similar to diffusion of heat. The same mathematics are obeyed, although the detailed mechanisms of heat and mass transport differ. A simple view of diffusion is that when a moisture sensitive material is placed into an environment containing a higher relative humidity, moisture travels through the material like a front. The depth to which the front has penetrated increases with the square root of time. This approximation allows rapid, although not highly accurate, estimation of conditioning times required for fiber and assemblies of fibers to reach equilibrium moisture content.

Accompanying absorption of water or any solvent are changes in the structural and physical properties of the fiber, including changes in mechanical properties, changes in electrical properties, an increase in the volume of the fiber, and a simultaneous release of heat. The volume changes are always greater radially than axially because of the molecular orientation in a fiber. The amount of heat released is generally proportional to the amount of moisture absorbed. Wool is capable of the largest thermal effects of all textile fibers. The total heat released in going from a dry fiber to a saturated fiber is the integral heat of sorption, plus the heat of condensation. The differential heat of sorption is a measure of the heat released in raising the moisture content of the fiber within only a narrow, specific range. Because the first molecules of moisture exposed to a dry fiber bind the tightest, more heat is released when adding moisture to a dry fiber than on adding the last molecules of water to saturation. Thus, the differential heat of sorption decreases with increasing moisture content or regain.

References

D. L. Allara and W. L. Hawkins, eds. *Stabilization and Degradation of Polymers*. Washington, DC: Americal Chemical Society, 1978.

J. Brandrup and E. H. Immergut, eds. *Polymer Handbook*, 3rd ed. New York: Wiley-Interscience, 1989.

T. F. Cooke. "Current Concepts on Superabsorbent Fibers." *INDA Journal of Nonwovens Research* 4 (1992), 41–51.

P. J. Flory. *Principles of Polymer Chemistry*. Ithaca, NY: Cornell University Press, 1953.

F. P. Incropera and D. P. DeWitt. *Fundamentals of Heat and Mass Transfer*, 3rd ed. New York: Wiley, 1990.

W. E. Morton and J. W. S. Hearle. *Physical Properties of Textile Fibers*, 3rd ed. Manchester, England: The Textile Institute, 1993.

F. Rodriguez. *Principles of Polymer Systems*, 3rd ed. New York: Hemisphere, 1989.

A. M. Schneider, B. N. Hoschke, and H. J. Goldsmid, *Textile Research Journal* 62 (1992), 61.

P. G. Shewmon. *Diffusion in Solids*. New York: McGraw-Hill, 1963.

Problems

(1) You purchase 1000 lbs of cotton at 20 °C, 95% RH, for $1000. About how much money were you overcharged if cotton costs $1/lb at 20 °C and 65% RH? Repeat the problem substituting wool for cotton. (R_{wool} at 20 °C and 95% RH is 27%.)

(2) If the moisture content of wool is 30%, what is its regain?

(3) Both neoprene and natural rubber are crosslinked rubbers. When these materials are subjected to "solvents," they swell, but the crosslinks prevent the polymer from dissolving. Sketch the amount of gasoline absorbed and desorbed as a function of vapor pressure of gasoline for neoprene and natural rubber. The absorption of gasoline at the saturation vapor pressure of gasoline is, say, 1600%, and that of neoprene is 2%.

(4) A fiber 25 μm in diameter requires 8 minutes for water to reach the center and equilibrate. How long will it take water to reach the center of a

(a) very closely packed yarn of 7 of the same fibers?

(b) very tightly packed yarn of 1000 fibers?

(c) loose yarn of 1000 fibers?

(5) A dry cotton fiber is placed on a vibrascope and tested with a hung weight of 200 mg. At time t the fiber in the vibrascope is exposed to air at 95% RH. Sketch fundamental frequency as a function of time. The frequency at time $< t$ is 1730 Hz.

(6) Discuss why the heat evolved when wool absorbs water is 1340 J/g at 0% RH and only 550 at 30% RH.

(7) It takes the fiber in problem 4 half a second to reach thermal equilibrium when plunged into an oven at 150 °C. How long will it take the bundles of fiber in 4a and 4b to reach equilibrium?

(8) Discuss the bonding associated with successively added water molecules in wool as the RH gradually increases from 0 to 100%.

(9) Explain why aqueous sodium hydroxide swells cotton.

(10) Which is larger, moisture regain or moisture content?

(11) At about what RH does "free water" begin to exist in hygroscopic fibers?

(12) What is meant by hysteresis in moisture regain?

(13) What is the nature of moisture pickup in PET, cited in Table 6.2?

(14) Per gram of water absorbed, does wool release more heat than other fibers?

(15) Give physical arguments to explain why the diffusion coefficient is a function of moisture content.

(16) Why does water swelling stop rather than proceed without limit to dissolution in cotton and wool?

(17) Calculate the percent increase in specific surface area associated with wetting a dry wool fiber.

(18) A wool fiber removed from a sweater shrinks 15% when exposed to boiling water. Why?

(19) Using the data in Table 6.1, select a potential solvent for
   (a) nylon.
   (b) acrylic.
   (c) PET.

(20) When moisture diffuses into a fiber such as wool or wood to change composition, what other changes occur in the fiber?

(21) An inexpensive hygrometer is constructed so that the axial expansion of a fiber leads to a meter reading. What fibers are good candidates for this application?

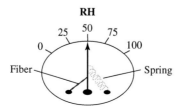

(22) Given that polymers contain covalent bonds, which are quite strong bonds, explain why most polymers degrade rapidly when exposed to ultraviolet light or sunlight.

(23) A fiber needs to be selected for use in a lightweight, flexible hose that is designed to carry liquid ammonia. The fiber will be braided and used to reinforce neoprene rubber. Discuss the attributes of rayon, nylon, Kevlar®, Spectra®, PET, and steel for the reinforcement material.

(24) Why do fibers typically swell more radially than axially when exposed to swelling agents such as water, as shown in Table 6.5?

(25) Surgeons routinely use fibers to repair wounds to the body. Internal sutures need to retain mechanical properties until the body has partially repaired itself, yet eventually the suture needs to be completely resorbed by the body. Discuss what sort of fiber chemistry is appropriate for a suture.

(26) Explain why wood and plywood warp when exposed to moisture gradients.

(27) How much longer will it take a 15 denier wool fiber to achieve moisture equilibrium than it will a 3 denier wool fiber? (Assume the density of wool is 1.31 g/cm^3, is invariant with RH, and the fiber is perfectly round and homogeneous.)

(28) One gram of water vapor is added to 1 kg of dry wool. How much heat is released? Compare the result with the value based on the addition of 1 g of liquid water added to dry wool.

(29) The regain of PET at 65% RH (room temp) is only 0.4%, and the heat of wetting is 5 J/g of fiber. Silk has a regain of 10%, and the heat of wetting is 69 J/g of fiber. Describe how the moisture is bound and what bond information you can deduce from the heat of wetting.

(30) What considerations are required for use of polymers in space?

(31) Despite its high cost, leather continues to be an important material for which there is no accepted substitute. What are the properties of leather that make it both so useful and difficult to reproduce?

(32) The kinetics and amount of moisture that intrudes into a fiber can be used as a structural probe. Consider exposure of Kevlar® to air with an increasing relative humidity. Discuss the interaction of moisture with Kevlar® using your knowledge of the chemical and physical structure of the fiber.

(33) A crosslinked polyisoprene rubber swells 6.2 times its original volume at 20 °C when immersed in a solvent with an interaction parameter of 0.39. Calculate the molecular weight between crosslinks, given that the molar volume of the solvent is 106 $cm^3/mole$ and the polymer density is about 1.0 g/cm^3.

PART TWO

Mechanical Properties

7

Tensile Properties

Tensile mechanical properties are one of the most important properties of fibers, particularly industrial fibers, since they are typically used in tension or complex stress states that include tension. It is imperative that you study this chapter closely and achieve a working knowledge of the terminology and concepts that are developed. In Chapter 8 we will study other important mechanical behavior, such as shear, torsion, and compression.

We begin our effort with basic definitions of stress and strain and the particular forms of stress historically used in the textile field. When reading some of the older literature in the area of mechanics of textile fibers, the name R. Meredith is often cited. He conducted some excellent basic research on the properties of textile fibers. A number of the figures and tables used in this chapter are adapted from his works.

7.1 Basic Definitions

Because of their long, slender form, fibers are natural candidates to be subjected to uniaxial tensile forces. A tensile force is one that leads to stretching of the material, as shown in Figure 7.1. Tensile stress is defined as the applied force or load normalized to the original cross-sectional area, F/A_o. Stress is used more often than force because the normalization facilitates comparison of samples with different cross-sectional shapes or diameters. We use the Greek letter σ to represent stress.

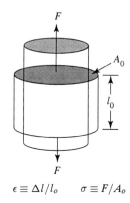

$$\epsilon \equiv \Delta l/l_o \qquad \sigma \equiv F/A_o$$

FIGURE 7.1 Tensile force on a fiber.

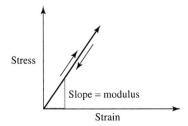

FIGURE 7.2 Stress–strain behavior in Hookean region.

One response of a fiber to a tensile force is extension of the material. When a fiber of original length l_o is stressed along its axis and extends an amount Δl, then the strain in the sample is $\Delta l/l_o$. We use the symbol ϵ to represent strain. A synonym for strain is elongation. Note that the units of stress are force per unit area. Strain is a dimensionless quantity, often reported in percentage.

When a material is subjected to a small stress, the stress causes the atoms to move from their original equilibrium positions in an elastic manner. When the stress is removed, the atoms return to their original positions and the sample recovers completely from the deformation. In this region of small stress, called the Hookean elastic region, stress and strain are linearly proportional, as shown in Figure 7.2. The proportionality constant between stress and strain is the slope of the line. It is called Young's modulus or the initial modulus. We will use the symbol E for Young's modulus. Being the slope of the stress–strain curve, Young's modulus is a measure of the amount of deformation that is caused by a fixed small amount of stress. Materials with high modulus, often called stiff or hard materials, deform or deflect very little in the presence of a stress, whereas materials with low modulus, often called soft materials, deflect significantly. An example of a low modulus material is a rubber band. High modulus material may be steel or ceramic, as sketched in Figure 7.3. Modulus is an important physical property of a material.

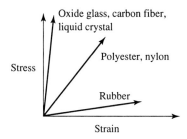

FIGURE 7.3 Initial modulus of various materials shown graphically.

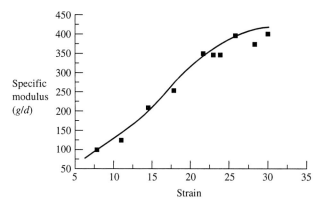

FIGURE 7.4 Effect of prior drawing on the modulus of polyethylene. (Note that a strain of 20 means the material was stretched 2000%, or 21 times its original length.)

At this time in most material texts, a table of moduli would appear; however, this text is largely about polymers. The modulus of natural fibers depends on molecular arrangement or orientation as well as on chemical structure. With increasing molecular orientation along the fiber axis, more molecules bear the load, and a decreasing deflection occurs at a given stress. Hence, modulus increases with molecular orientation. The modulus of a synthetic fiber depends on precisely the same factors; however, man has tremendous control over molecular orientation. Hence, the modulus of most synthetic fibers will vary according to the structure of the fiber that develops during processing. To achieve the high modulus and strength required of industrial fibers, elaborate steps are usually taken in processing to stretch out the molecules as much as possible. Some industrial fibers may be drawn in several separate steps at various temperatures and rates to achieve the high molecular orientation required to give high modulus, as shown in Figure 7.4.

The cost of the high modulus and strength is low elongation-to-break, typically less than 20 percent for industrial PET or nylon fibers. Liquid crystalline fibers represent perhaps the ultimate in molecular orientation. The molecules are rigid, and they align almost parallel to the fiber axis. The inherent molec-

TABLE 7.1 Modulus of Kevlar® Fibers

Fiber	Modulus (GPa)	Specific Modulus (g/d)
Kevlar® 29	50	420
Kevlar® 49	125	980
Kevlar® 149	185	1450

Source: "Preliminary Engineering Data," DuPont Corp., E-95614.

ular stiffness plus the high molecular orientation in fibers give LCP's extremely high moduli. Still, various grades of liquid crystal fiber are available with different moduli and strain-to-fail, depending on chemical composition, and the molecular orientation imparted, often by using hot stretching, as illustrated in Table 7.1. Note the units of modulus most commonly used are GPa, gigapascals, which is 10^9 N/m^2.

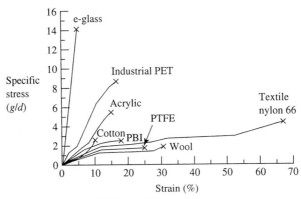

(a) Tensile behavior of textile fibers

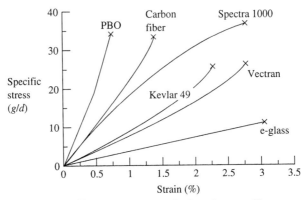

(b) Tensile behavior of high performance fibers

FIGURE 7.5 Stress–strain behavior of various fibers.

TABLE 7.2 Typical Tensile Properties of Fibers

Fiber	Specific Modulus (g/d)	Strain-to-Fail (%)	Tenacity (g/d)
Natural			
Cotton	60	7	4
Wool	25	40	1.6
Silk	80	23	4
Viscose rayon	68	18	2.5
Flax	200	3	6
Synthetic			
Nylon	35	25	5.5
PET	130	10	6
Acrylic	70	25	4
PP	80	17	7
Polyurethane	0.1	500	0.4
Kevlar® 149	1450	2	35
s-glass	410	2	20
T-1000 C fiber	1800	2	44

Sources: R. Meredith, *Journal of the Textile Institute* 36 (1945), T107; S. Kumar, *Indian Journal of Fiber and Textile Research* 16 (1991), 52.

Let us return to Figure 7.1. These curves represent the initial portion of a stress–strain curve. We need to consider now what occurs when stress or strain increases without limit. Eventually, at some stress, the material will fail. With oxide glass, carbon, ceramic, and other brittle fibers, failure occurs in the Hookean elastic region. We indicate failure by placing a × at the termination of the σ–ϵ curve. Most textile fibers extend into a nonlinear region prior to failure, as shown in Figure 7.5. The nonlinear behavior characteristic of most of the σ–ϵ curves for textile fibers is indicative of the fact that the atoms are no longer strained within their elastic limit, but rather, atoms and sections of molecules are beginning to move past one another. Such deformation, in general, is not recoverable. It is called plastic deformation. In polymers, plastic deformation tends to align the molecules along the stress direction. This reorientation will occur up to the point of failure. Typically, failure is initiated at a local point of weakness. The stress at failure is called strength. Strength divided by density is specific strength, or tenacity.

Modulus and strength are key concepts used over and over in this text. It is important to recognize an essential difference between the two terms. Modulus is an intrinsic property. It depends on bulk material behavior. Strength, on the other hand, is an extrinsic property. It depends on only a small portion of the material, the weakest portion. For this reason, the modulus provides information on the bulk structure of a material. Getting high-strength fiber first requires high molecular orientation; then it requires elimination of all strength-limiting defects. Approximate values for the tensile mechanical properties of various textile and industrial fibers are shown in Table 7.2.

A few more commonly used terms need to be defined. Toughness, or work-to-break, is the area under the stress–strain curve. This makes good sense, since work is force exerted through a distance. Toughness is a measure of the amount of work required to cause failure. Brittle materials are not tough, nor are very low modulus materials. Materials with reasonable values of strength and strain-to-fail have good toughness. In the realm of polymers, tough materials are usually semicrystalline. Glassy polymers are brittle, and rubbers have low modulus. Textile fibers need to be tough. Hence, most textile fibers are semicrystalline, a fact we learned in Chapter 3.

7.2 True Stress, Specific Stress, Tenacity, and Breaking Length

The stress defined in the previous section, force per unit of original cross-sectional area, is known as engineering stress. There are, however, many other ways to represent stress. We need to understand a few of these ways. The first is true stress. True stress is designed to show the actual stress a material is subjected to at any point along the stress–strain curve. It is especially useful when a material can withstand large strains. Consider a fiber with the stress–strain curve shown in Figure 7.6. Let us examine one point on the stress–strain curve, that at $\epsilon = 4.0$. At this point the engineering specific stress is 0.4 g/d, but the sample has increased in length 400 percent. Hence, the cross-sectional area is 1/5, or $1/(1 + \epsilon)$, times the initial value. True stress is defined as force normalized to the instantaneous cross-sectional area, F/A. At the point of interest here, true stress, σ_T, is five times the engineering stress. We can rapidly derive an expression for true stress based on engineering stress and strain:

$$\sigma_T = F/A = (F/A_o)(A_o/A) = \sigma(A_o/A)$$

We note that when a material deforms outside the Hookean elastic region, volume is conserved:

$$V = A_o \times l_o = A \times l$$

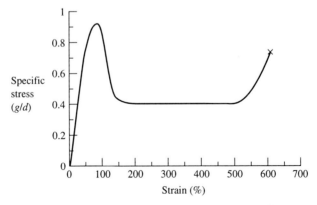

FIGURE 7.6 Stress–strain behavior of unoriented polypropylene.

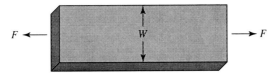

FIGURE 7.7 Tensile testing of nonwoven materials.

where, as usual, A and l are instantaneous values of cross-sectional area and length. Substitution of $A_o/A = l/l_o$ gives

$$\sigma_T = \sigma(l/l_o) = \sigma(l_o + \Delta l)/l_o = \sigma(1 + \Delta l/l_o)$$

$$= \sigma(1 + \epsilon)$$

This expression is exactly what we anticipated. The true stress is the engineering stress corrected for the reduction in cross-sectional area. The reduction in cross-sectional area is the stretch ratio,

$$\lambda = 1 + \epsilon$$

The discussion on mechanical properties to this point has been general. Stress and strain are what most scientists and engineers who study mechanics use. Unfortunately, it is not the system that is commonly used by textile and fiber scientists, paper scientists, composite scientists, or nonwoven scientists. They use the related properties, specific stress and specific modulus, which are stress and modulus divided by density. Unfortunately, scientists in each of these fields express specific strength and specific modulus differently, as you will see shortly.

Fiber scientists express stress and strength in units of grams-to-fail normalized to the denier of the fiber or yarn, g/d or gpd. Since denier has units of g/9000m, g/d has units of 9000 m, or length. (The unit of g/tex is similar. Only the numerical constant differs, 1000 rather than 9000.) We will return to the significance of this length in just a minute.

Composite scientists are usually concerned with strength normalized to the weight of the part, so they use stress normalized to density. The units of stress are, in the U.S. system, lbs/in^2. The units of density are lbs/in^3. Hence, the units of σ/ρ are inches. Specific stress has units of length.

Nonwoven and paper scientists are concerned with measuring the thickness of their material, since stress is force per thickness per width. Determination of thickness is a nontrivial issue, since the precise thickness value measured depends on how much the measuring device squeezes the material. Thus, most scientists use basis weight rather than cross-sectional area. Basis weight is the mass of a unit facial area of material. The units of specific strength are force divided by basis weight divided by the width of the test material, as shown in Figure 7.7:

$$\text{Specific strength} = F/(W \times BW)$$

where

F = load-at-failure,

W = width of test piece, and

BW = basis weight = mass per unit facial area.

The units are lbs/[(lbs/in^2)(in)] = inches. Paper scientists call this length break-ing length. It represents the maximum length of a sample of fiber, paper, nonwoven, or composite that could be let out from the top of the Empire State Building before it would break under its own weight. Breaking length, then, is not a measure of strength, but it is a measure of specific strength. The unit g/d is breaking length in meters/9000.

The term of specific strength used by fiber scientists is tenacity. It is the maximum specific stress a fiber can bear. Often tenacity and strength are used interchangeably. This is technically incorrect, and we shall try to avoid doing so.

Modulus has the same units as stress. Specific modulus is modulus divided by density. The units for specific modulus are the same as those for specific stress. Fiber scientists usually use the units g/d or g/tex for specific modulus. The modulus and strength of oriented polymers may meet or exceed those of metals or ceramics when normalized to density, a fact that you may observe in various tables throughout this text.

7.3 Elastic and Plastic Deformation, Hooke's Law, and Poisson's Ratio

In the previous section we showed that the beginning of a stress–strain curve has a linear portion. In this region Hooke's law is obeyed:

$$\sigma = E\epsilon$$

Outside this region Hooke's law is not obeyed. Outside Hooke's region the deformation done to the atoms and molecules is generally not recoverable. This does not mean that the covalent bonds necessarily break, but rather that the long chain molecules slip past one another, often never to return. The slip process improves molecular orientation along the fiber axis. Hence, if a poorly oriented fiber is stretched and then tested in a testing machine, the poorly oriented fiber and the stretched fiber have different stress–strain behavior, as shown in Figure 7.8. Indeed, improving the molecular orientation is one tar-get of synthetic fiber processing, in both extrusion, where the take-up speed exceeds the extrusion speed, and post-extrusion stretching. The stretching pro-cess can be simulated on a tensile testing machine, especially if the machine can be fit with an oven capable of reaching the appropriate stretching temper-ature. For example, if the most stretchable fiber in Figure 7.8 is extended 550 percent, then the new curve for that fiber resembles that of the most highly oriented fiber in Figure 7.8.

Returning to Figure 7.8, we note that the metamorphosis of mechanical be-havior with molecular orientation is clearly shown. Both modulus and strength

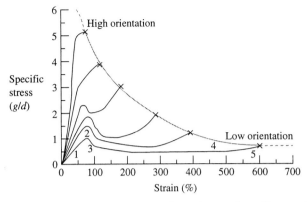

FIGURE 7.8   Stress–strain curves of a polypropylene fiber series.

increase with molecular orientation. The strain-to-fail decreases with molecular orientation, since the chains are no longer coiled on themselves, but rather, oriented chiefly in the fiber direction. We examine the stress–strain curve for the unoriented fiber and note several specific regions:

1. The initial Hookean region.
2. The yield point, which is local maximum in the curve.
3. Neck formation and propagation, during which time a local region of area reduction forms and propagates to the ends of the specimen.
4. Reinforcement, which is uniform molecular alignment.
5. Failure.

Neck propagation and molecular orientation can be readily observed by drawing an unoriented film of polyethylene between crossed polars on a transparency projector. A photomicrograph of a neck in PET fiber is shown in Figure 7.9. Steps 2–4 are characteristic of only poorly oriented samples. They do not occur in oriented fibers, as shown in Figure 7.8.

Although the linear region of the stress–strain curve may be small relative to the entire deformation curve, it is nonetheless important. Poisson's ratio is another important material characteristic that deals with the behavior in the region of Hooke's law. Poisson's ratio, ν, is defined as the ratio of lateral contraction to axial extension under tensile deformation. The accounting is done using strains:

$$\nu = -\epsilon_x / \epsilon_z$$

where

ϵ_x is the strain in the lateral direction, $\Delta x / x_o$, and
ϵ_z is the strain in the axial direction, $\Delta l / l_o$.

Note that ϵ_x is negative, so that Poisson's ratio is positive. Poisson's ratio must be between 0 and 0.5. It is a measure of the volume change associated with deformation in the Hookean elastic region. When $\nu = 0.5$, there is no volume

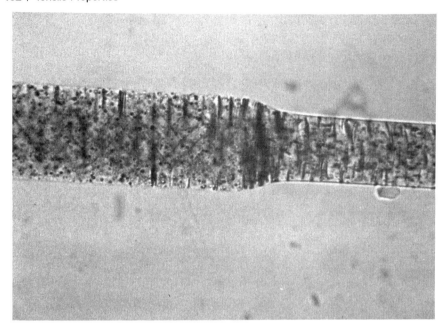

FIGURE 7.9 Photomicrograph of neck in stretched poly(ethylene terephthalate) fiber.

change during tensile deformation. The material is fluidlike. Most polymers have a Poisson's ratio of 0.3 to 0.4. Metals and ceramics range from 0.20 to 0.35. Poisson's ratio, like Young's modulus, is an important material property. We will use Poisson's ratio later in the text for various calculations.

7.4 Stress–Strain Curves

So far in this chapter we have discussed in general the nature of tensile deformation and stress–strain curves of fibers. The effect of molecular orientation on modulus, strength, and strain-to-fail has been addressed. In this section we will investigate the effect of helix angle on mechanical behavior, the effect of moisture and other fluids on tensile properties, and present some basic concepts associated with a form of elasticity unique to polymers, rubber elasticity.

7.4.1 The Effect of Helix Angle

The effect of helix angle on modulus, strength, and strain-to-fail is an important consideration in fiber science. In natural fibers the long molecules are helically oriented about the fiber axis, as described in Chapter 3. In addition, yarns are a collection of fibers. Most yarns are twisted, so the fiber assumes a helical path. Hence, the calculations associated with helix angle are applicable to natural fibers as well as to twisted yarns and ropes. Molecules in synthetic fibers may or may not have helical conformations, depending on steric

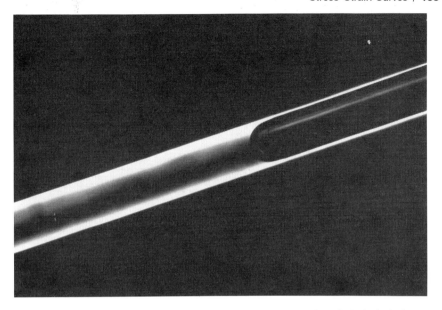

FIGURE 7.10 Optical photomicrograph of milkweed fiber with air bubble in lumen.

hindrance and polarity of the side groups. Poly(butylene terephthalate) and polyacrylonitrile have helical chains.

A fiber with molecules arranged in a helical path may have reasonable mechanical properties both along and normal to the fiber axis. Consider, for example, a fiber from the milkweed seed head, as shown in Figure 7.10. These fibers need to be light, so as to keep the seed airborne. Thus, nature has engineered and manufactured a hollow, thin-walled fiber in which the molecules have a helical orientation. Because of the biaxial orientation of molecules in the fiber, the tube does not collapse.

The effect of helix angle, θ, on axial mechanical properties can be calculated with the aid of Figure 7.11. We solve the problem for a fiber that is part of a twisted yarn. Let us first consider the strain in a helical fiber stretched

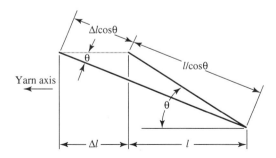

FIGURE 7.11 Fiber with helical path.

along the yarn axis. The strain imposed on the yarn by an external force is

$$\epsilon_{\text{yarn}} = \Delta l / l$$

We can relate the strain in the yarn to that in the fiber with helix angle θ:

$$\Delta l_{\text{fiber}} = \Delta l \cos\theta \qquad \text{and} \qquad l_{\text{fiber}} = l / \cos\theta$$

Substitution gives

$$\epsilon_{\text{fiber}} = \Delta l_{\text{fiber}} / l_{\text{fiber}} = \Delta l \cos\theta / (l / \cos\theta) = \epsilon_{\text{yarn}} \cos^2\theta$$

This relationship is usually expressed in the opposite way, showing how the yarn strain varies with the fiber strain:

$$\epsilon_{\text{yarn}} = \epsilon_{\text{fiber}} / \cos^2\theta$$

The strain on the fiber is reduced from that on the yarn by $\cos^2\theta$. Similar analysis can be used to show the relationship between the stress on the fiber and that on the yarn. Here a $\cos\theta$ develops for the resolution of force and another $\cos\theta$ for the resolution of fiber cross-sectional area:

$$\sigma_{\text{yarn}} = \sigma_{\text{fiber}} \cos^2\theta$$

Combining these two expressions gives us one for the effect of helix angle on modulus:

$$E_{\text{yarn}} = \sigma / \epsilon = \sigma_{\text{fiber}} \cos^2\theta / (\epsilon_{\text{fiber}} / \cos^2\theta) = (\sigma / \epsilon) \cos^4\theta = E_{\text{fiber}} \cos^4\theta$$

Thus, the modulus of a yarn consisting of fibers with a uniform helix angle θ is reduced by $\cos^4\theta$, the strain-to-fail is increased by $\cos^2\theta$, and the strength is reduced by $\cos^2\theta$. On this basis we expect

- Natural fibers to have a lower modulus than that of the molecules from which they are made.
- Yarns to have a modulus that decreases and strain-to-fail that increases with increasing twist.

Such considerations often have practical applications. Consider, for example, the behavior of a 10 dpf liquid crystal fiber proposed for use in tire cord. This LCP fiber is characterized by a strain-to-fail of only 0.6 percent in compression, yet the strain imposed by the tire on an untwisted fiber is roughly 1.25 percent. Placing an untwisted yarn in the tire leads to immediate fiber failure. In addition, the act of twisting causes compressional failure of the fiber before it ever gets into the tire, as you shall see in the following chapter. Reducing the fiber denier, however, allows the yarn to be twisted. The helical path of the fiber in the rubber now reduces the imposed strain on the fiber in the rubber. Thus, the liquid crystal fiber denier was reduced to about 3 and the yarn was twisted to enable its use in tire cords.

7.4.2 The Effects of Moisture and Temperature

In Chapter 6 we learned that moisture diffuses into the bulk of fibers that are capable of hydrogen bonding. The water plasticizes the fiber, causing a soft-

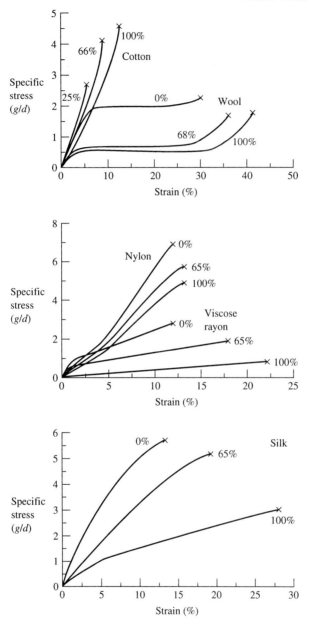

FIGURE 7.12 Effect of RH on tensile behavior of fibers. (R. Meredith in J. M. Preston, ed., *Fiber Science*, 2nd ed., Manchester, England: Textile Institute, 1953.)

ening or, more technically, a reduction in modulus. Strain-to-fail and strength are also affected, as shown in Figure 7.12. Cotton shows unusual, difficult-to-explain behavior, in that the strength increases with humidity. It has been suggested that cotton fiber has a high level of residual stress. Moisture allows the residual stresses to relax, and hence, tensile properties improve with moisture content. The theory has not been confirmed with experimental data.

The effect of temperature is similar to the effect of moisture on the mechanical properties of fibers, as shown in Figure 7.13. Heat and moisture each cause a softening of the noncrystalline regions, which reminds us of the similarity between dissolution and melting. In the case of PET, the data show that the modulus is nearly constant in the range of −57 to 99 °C, then drops substantially and remains almost constant between 150 and 205 °C. Between 99 °C and 150 °C is the glass transition temperature, where polymer segments on the length scale of a mer become mobile. The data for nylon are quite similar to those of PET.

When we take the data from Figure 7.13, or similar data, and plot initial or Young's modulus as a function of temperature, we get the sort of plot shown in Figure 7.14. The modulus is constant with temperature below about 70 °C, the glass transition temperature. At the glass transition temperature, the modulus decreases an amount inversely proportional to the crystallinity in the sample. Above the glass transition temperature, the modulus is essentially constant until the crystals melt. This behavior is characteristic of semicrystalline polymers, as shown in Figure 7.15. A noncrystalline polymer will show a large drop in modulus at T_g, as shown in Figures 7.15 and 7.16. If the noncrystalline polymer is crosslinked, then the modulus does not fall to zero in the vicinity of the glass transition temperature; rather, the modulus maintains a value that depends on the level of crosslinking up to the decomposition temperature. The region of the modulus curve above the glass transition temperature for crosslinked amorphous polymers is called the rubbery plateau. When the rubbery plateau is at or near room temperature, the polymer is an elastomer or a rubber, like vulcanized natural rubber. In the next section we will come to understand the room temperature stress–strain behavior of elastomers.

7.4.3 Basic Rubber Elasticity

A pure metal consists of identical atoms that act largely independently of one another. Ceramics are a combination of a metal and a nonmetal. Again, ions move more-or-less independently of one another. Polymers, on the other hand, are unique materials in that they are composed of long chain molecules. Atoms are covalently bonded along the polymer backbone. The unique structure provides polymers with some unique properties, such as rubber elasticity. While all materials are capable of Hookean elasticity, only rubbers can completely recover from extremely large deformations. A simple rubber band serves as an example. It can be stretched and restretched time and time again, showing virtually complete recovery between stretchings.

As mentioned earlier in this chapter, all materials can recover from Hookean deformations, since only elastic bond stretching and bond flexing are involved. How is it that elastomers, lightly crosslinked amorphous polymers above their glass transition temperature, can recover from high deformations? The key to the answer lies in the long chain nature of the material and the

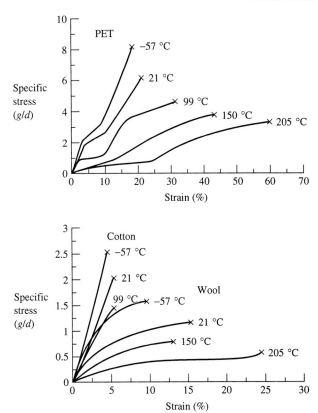

FIGURE 7.13 Effect of temperature on the mechanical properties of various fibers. (Technical Bulletin X-82, Wilmington, DE. Copyright 1958, E. I. du Pont de Nemours & Co., Inc.)

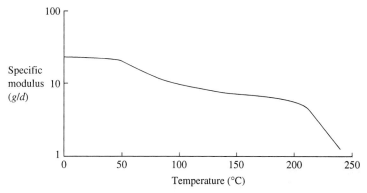

FIGURE 7.14 Effect of temperature on the modulus of modified Vectra® polymer.

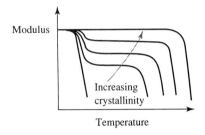

FIGURE 7.15   Effect of crystallinity on the temperature dependence of modulus.

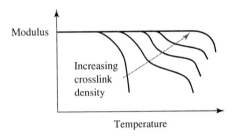

FIGURE 7.16   Effect of crosslink density on the modulus of a glassy polymer.

mobility associated with the chains above T_g. Elastomers consist of entangled chains that are coiled upon themselves. When the material is stretched, the chains uncoil and orient. The deformation remains within the elastic limit of each covalent bond even at high strains. Hence, when the stress is removed, the chains seek to return to their low energy coiled conformation, as illustrated in Figure 7.17. The crosslinks (or hard segments in the case of thermoplastic elastomers) prevent the chains from irrecoverably sliding past one another and give the chains a memory of their lowest energy state. The elastomeric polymer recovers almost completely from several hundred percent strain.

The mechanisms described in the preceding paragraphs are the essential features of rubber elasticity. The driving force for recovery is called maximization of entropy and is founded in thermodynamics. In other courses you will work through the thermodynamics of rubber elasticity. It is a particularly satisfying exercise, since the theoretical predictions of statistical thermodynamics are borne out in practice. The results of the mathematics of rubber elasticity show

$$\sigma = NRT(\lambda - \lambda^{-2})$$

where

σ = stress in the rubber,

N = the effective number of moles of polymer chains per unit volume,

R = gas constant,

T = absolute temperature, and

$\lambda = \epsilon + 1$ = stretch ratio.

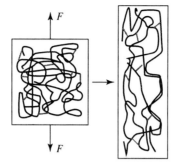

FIGURE 7.17 Molecular schematic of rubber elasticity.

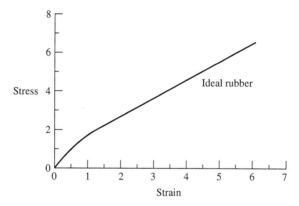

FIGURE 7.18 Stress–strain behavior of ideal rubbery material.

From this equation you can see that stress is not linearly proportional to strain, as is true for Hookean elasticity. The effect of strain on stress for an ideal elastomer is shown graphically in Figure 7.18. There are two chief reasons for presenting the basic concepts of rubber elasticity:

1. Some textile fibers are rubber elastic, such as the thermoplastic elastomers and synthetic and natural thermoset rubbers.

2. Rubber elastic behavior creeps into the stress–strain curves of a number of fibers, such as that of wool. Note the curvature in certain regions of the σ–ϵ curve for wool, shown in Figure 7.11. Wool has cysteine bonds (crosslinks) and has reasonable elastic recovery from moderate deformation, as we shall see in the following section. Note too that wet wool has enhanced molecular mobility in noncrystalline regions.

3. Most semicrystalline textile fibers show limited shrinkage when heated. The driving force for the shrinkage is entropy maximization.

In sum, rubber elasticity requires a continuous phase of noncrystalline polymer capable of fluidlike motion. The restoring force is based on entropy considerations and requires the molecules to have a memory of their original conformations.

7.5 Elastic Recovery

Elastic recovery is the ability of a material to recover from deformation. We learn about elastic recovery by conducting cyclic testing. A multi-cycle stretch and recovery plot is shown in Figure 7.19. In polymers the total elastic recovery is the sum of Hookean-based recovery and rubber-based recovery. The Hookean part is that which is stored in bond energy. The rubber part is that which is stored in molecular extension or entropy. The recoverable strain due to Hookean elasticity is linear, and that due to rubber elasticity is nonlinear.

7.5.1 Strain Recovery

Elastic recovery is the recovered (elastic) strain divided by the total strain imparted (elastic plus plastic) in the stretch cycle:

$$ER = \text{recovered strain/imparted strain}$$

The elastic recovery curves plotted as a function of strain are shown in Figure 7.20 for a number of fibers. Polypropylene, nylon, and wool have the highest elastic recovery from small and moderate deformations. Many applications of textile fibers require reasonable elastic recovery; however, elastic recovery of a fabric also depends on fabric construction. Knit fabrics have good stretch and recovery, largely because the yarn path is highly convoluted. One application in which elastic recovery is of critical importance is carpet fiber. Nylon dominates the synthetic carpet industry. Wool is the classic natural carpet fiber.

Moisture affects the ability of a hygroscopic fiber to recover from deformation, as shown in Table 7.3. The data show that moisture influences elastic recovery, but the effects are complex and no obvious conclusions may be drawn.

When a natural fiber has been structurally altered by mechanical straining into the plastic region, the conditioned fiber has a higher modulus, lower strain-to-fail, and higher strength; however, the original mechanical properties of hygroscopic fibers may be largely restored by treating the fiber with heat, steam, or hot water. The water plasticizes the noncrystalline regions,

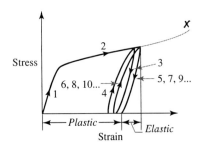

FIGURE 7.19 Stretch and recovery curves of a typical fiber.

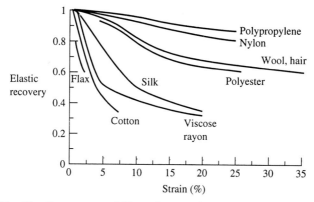

FIGURE 7.20 Elastic recovery of fibers from various imposed strains. (R. Meredith, *Journal of the Textile Institute* 36 (1945), T147, and G. Cook, *Handbook of Polyolefin Fibers*, Manchester: Merrow, 1968.)

enhances mobility, and reduces T_g. This may be a blessing, if such effects are desirable. On the other hand, it may be a curse, for example, in a finished garment, since shrinkage may accompany structural restoration. When mechanically conditioned wool is treated with steam or hot water to restore properties, the process is appropriately called swelling recovery.

7.5.2 Work Recovery

Another useful measure of recovery is the work recovery from tensile deformation, which again is a fractional quantity:

$$WR = \text{work recovered/work done in stretching}$$

Work (or toughness) is the area under the load–extension curve. The area under the stress–strain curve is work per unit volume. (The units of the area

TABLE 7.3   Effect of Moisture on Elastic Recovery

| | Elastic Recovery from Tensile Strain of Fiber | | | | | |
| | 1% Strain | | 5% Strain | | 10% Strain | |
	60% RH	90% RH	60% RH	90% RH	60% RH	90% RH
Natural						
Cotton	91	83	52	59		
Wool	99	94	69	82	51	56
Viscose	67	60	32	28	23	27
Silk	84	78	52	58		
Synthetic						
Nylon	90	92	89	90	89	
PET	98	92	65	60	51	47
Acrylic	92	90	50	48	43	39

Source: L. F. Beste and R. M. Hoffman, *Textile Research Journal* 20 (1950), 441.

under the load–extension curve might be in pounds, whereas those under the stress–strain curve might be lbs/in^2 = in-lbs/in^3.) Work not recovered is a measure of the work lost to the material, usually in the form of heat. When conducting these sorts of tests on rubbery materials, hysteresis is often mentioned. The work loss associated with repeated deformation is the area within the hysteresis loop, as shown in Figure 7.19. That is, the area between the two curves—stretch and recovery—represents the work lost to the material. Generally, most lost work is converted to heat. Heat development is usually a detrimental property in rubbers subjected to cyclic loading. This is in part due to the fact that so much heat can be generated that the material begins to degrade. An example of an application where mechanical hysteresis from deformation is critical is in an automobile tire. Today's automobile tires, which have a rather low work loss, heat up to an acceptable, albeit perhaps surprising, 125 °C on the highway. Even at these temperatures, chemical reactions such as hydrolysis may occur at a significant rate in the tire, which can lead to degradation of organic tire cords and interfaces. In Chapter 10 we study dynamic mechanical properties. Chapter 10 includes a discussion of how heat is generated in polymers by a process called internal friction, which is one mechanism responsible for hysteresis.

7.6 Summary

Fiber mechanics is the study of the mechanical properties of fibers and fibrous assemblies. The tensile properties of fibers are key to their selection and success in many applications. Normalized load is called stress, or engineering stress, and normalized extension is called strain or elongation. The stress–strain curve shows many of the important mechanical characteristics of fibers. The initial slope of the stress–strain curve is the stiffness, or modulus, of the material; the stress-at-failure gives the strength; and the strain-at-failure gives the elongation-to-break. The initial linear part of a stress–strain curve is a result of bond stretching and flexing, and it is completely reversible. This region characterizes Hookean elastic deformation. When deformation becomes nonlinear, it is usually plastic. Plastic deformation is a result of molecules sliding past one another. In rubbers with the appropriate chemical and physical structure, called elastomers, some or all of the nonlinear deformation is immediately recovered upon removal of the stress. Such behavior is called rubber elastic.

Engineering stress is load divided by original cross-sectional area. True stress is load divided by the cross-sectional area of the sample at any stress or strain of interest. True strength provides an estimate of how strong a fiber would be if it were drawn to its maximum.

The modulus and strength of a fiber generally increase with molecular orientation. Elongation-to-break decreases with orientation. When the molecular or fiber path is oblique to the fiber axis, stresses and strains need to be resolved along the tensile or fiber axis. In the special case of a helical path, the effect of helix angle on yarn properties is to increase the apparent strain by $\cos^2$ of the helix angle; reduce stress and strength by the same factor; and consequently, reduce modulus by $\cos^4$ of the helix angle.

Moisture affects the mechanical properties of hygroscopic fibers. Basically, the moisture lodges in the noncrystalline regions and plasticizes them, reducing the modulus.

Rubber elasticity describes the unique mechanical behavior of noncrystalline materials that are above their glass transition temperature. Long chain molecules seek to coil on themselves when crystal formation is not favorable. When the fluidlike molecules are stretched out and the force is removed, the molecules return to their original positions as long as they had been imparted a memory by crosslinking or otherwise. Consequently, elastomers typically show high elastic recovery, whereas semicrystalline fibers generally show moderate to poor elastic recovery. Of the semicrystalline fibers, nylon, polypropylene, and wool show the best elastic recovery.

References

J. W. S. Hearle, P. Grosberg, and S. Backer. *Structural Mechanics of Fibers, Yarns, and Fabrics*, V1. New York: Wiley-Interscience, 1969.

W. D. Kingery, H. K. Bowen, and D. R. Uhlmann. *Introduction to Ceramics*, 2nd ed. New York: Wiley-Interscience, 1976.

W. E. Morton and J. W. S. Hearle. *Physical Properties of Textile Fibers*, 3rd ed. Manchester, England: The Textile Institute, 1993.

J. P. Schaffer, A. Saxena, S. D. Antolovich, S. B. Warner, and T. H. Sanders, *Materials Engineering*. New York: Times Mirror Books, 1995.

I. M. Ward. *Mechanical Properties of Solid Polymers*, 2nd ed. New York: J. Wiley, 1983.

I. M. Ward, ed. *Structure and Properties of Oriented Polymers*. Essex, England: Applied Science Publishers, 1975.

Problems

(1) Develop a formula to facilitate conversion from giga or mega Pascals to g/d.

(2) Referring to the following table as needed, sketch the $\sigma-\epsilon$ curves on a single set of axes for oxide glass, staple polyester, and elastomeric polyurethane (properly scaled and everything labeled).

Fiber	Diameter	Density	Test Length	Load-at-Failure	Extension
Oxide glass	10 μm	2.5 g/cm^3	2 cm	30 g	1.2 mm
Urethane	20 μm	1.1 g/cm^3	2 cm	3 g	0.12 m
Staple PET	16 μm	1.35 g/cm^3	2 cm	30 g	1.2 cm

(3) In class we discussed that an oriented fiber is made by drawing and "walking up" the stress–strain curve to the desired point. Sometimes, however, it is desirable to be able to "walk down" a stress–strain curve to reduce the modulus and increase the strain-to-failure. How might this be achieved, or must the sample be remelted, re-extruded, and reprocessed?

(4) The load–extension curve shown was obtained from a 2.2 denier cotton single filament using a 5 cm gauge length. Calculate tenacity, true tenacity, specific modulus, load at 3% strain, and initial cross-sectional area ($\rho = 1.52$ g/cm^3).

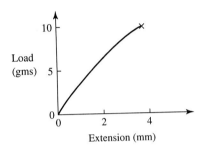

(5) Discuss the significance of the glass transition temperature for semicrystalline fibers and elastomeric fibers. Specifically, state in general terms what occurs at a T_g. Explain how T_g influences mechanical properties, especially modulus.

(6) For the fiber whose load–extension curve is shown, calculate specific stress in g/d at 10% strain and initial specific modulus in g/d. The initial gauge length is 1 cm and the fiber denier is 3.0.

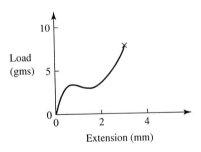

(7) Derive the equality $\sigma_T = \sigma(1 + \epsilon)$ using the definitions of true and engineering stress and the observation that volume is essentially conserved.

(8) Derive the result that Poisson's ratio of a fluid or any material that shows no volume change when subjected to a tensile deformation is 1/2.

(9) The following table summarizes some of the properties of cotton and wool. Use your knowledge of structure to explain the differences in properties.

Wool		Cotton
25	specific modulus (g/d)	60
40	strain (%)	7
1.6	tenacity (g/d)	4
0.99	elastic recovery from $\epsilon = 2\%$	0.75

(10) Suggest why the modulus of PET is greater than that of PBT [poly(butylene terephthalate)]. Why is the elastic recovery of PBT better than that of PET? (Hint: Consider crystal structure and the analogy to wool.)

(11) An unoriented polymer is stretched to the point indicated on its stress–strain curve below. The sample is removed, remeasured, and regripped in the testing machine. Estimate the strain-to-fail and tenacity (g/d) of the sample.

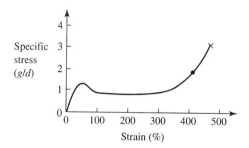

(12) Calculate the specific modulus, tenacity, true tenacity, work-to-break, and strain-to-fail for the fiber shown above.

(13) Describe hysteresis in stretch and recovery. Where does the "lost" energy go? What happens when a noncrystallizing silicone rubber band is repeatedly stretched and relaxed in an insulated small volume of water?

(14) A friend does a series of tests and informs you that every material he tests has 100% elastic recovery, including a variety of fibers, such as PE, PET, etc. How could he be correct?

(15) Show the $\sigma-\epsilon$ cycles for a sample of unoriented PE fiber that is stretched to 70% of its strain-to-fail, the stress removed, and the process repeated two more times.

(16) What is swelling recovery? Why does it apply only to wool?

(17) The load–extension curves of a series of 3 denier/filament PET tested in uniaxial tension using a gauge length of 2 cm are shown here:

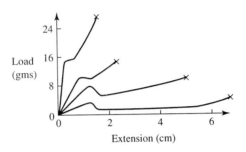

Calculate
(a) specific strength or tenacity (g/d).
(b) initial (specific) modulus.
(c) toughness.
(d) strain-at-failure.

(18) Calculate the true tenacity of the fibers in the preceding problem. Comment on the similarities of the values for the samples.

(19) Show with a series of sketches how molecular orientation influences the shape of a stress–strain curve.

(20) The yield point may be determined in any one of a variety of arbitrary ways. The critical issue is: What occurs at the yield point?

(21) Why are very high tenacity fibers in general *not* used in apparel applications? For example, PET can be made to a tenacity of 10 g/d, but it is used in reinforcement applications only.

(22) Using the data in problem 9 and the appropriate densities, calculate the strength and modulus of cotton and wool.

(23) Calculate the strength, initial modulus, strain-at-failure, and work-to-break for each of the following materials (gauge length = 5 cm, fiber diameter = 20 μm).

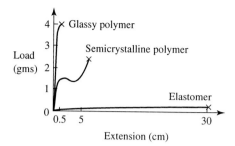

(24) Discuss how the mechanical behavior of cotton differs from that of other polymers in the presence of moisture. Suggest a plausible reason for the behavior.

(25) What are the fundamental reasons for elastic recovery in materials, specifically polymers, especially textile fibers?

(26) Describe qualitatively how the initial modulus of cotton and polyester changes in the first five cycles of a tensile fatigue test.

(27) Use your knowledge of fiber structure to predict the $\sigma-\epsilon$ curves of superabsorbent fiber both dry and swollen to equilibrium in water. Plot the behavior of a wood pulp fiber on the same axis.

(28) List two applications for which a polymer fiber needs
   (a) high elastic recovery.
   (b) low modulus.
   (c) high work-to-break.

(29) Convert the following stress–strain curve of a single filament to a load–extension curve. The fiber denier is 3.0, density 1.38 g/cm^3, and the test length is 5 cm.

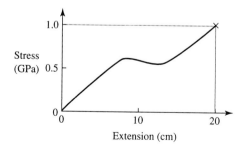

(30) Using Table 7.3 as a reference, rationalize the values of recovery from 5 and 10% tensile strain for cotton and wool fibers.

(31) In the chapter it was mentioned that PP, nylon, and wool have the best recovery from deformation. Wool and nylon dominate the carpet industry. Why is PP not more extensively used, since it has better elastic recovery than either nylon or wool?

(32) You are able to extend the molecules of PE along the fiber axis, as is done in a fiber such as Spectra® 1000. Why does the polymer not shrink as soon as the extension force is removed? Plot how the length of a 10 cm section of fiber changes as a function of increasing temperature.

8

Mechanical Properties of Fibers: Shear, Bending, Torsion, and Compression

When materials are deformed or strained, stresses develop. The study of the relationships between stresses, strains, and material properties is called mechanics. We have discussed a number of important concepts associated with tensile properties of fibers in the text. The importance of tensile properties is rather obvious. Consider that without sufficient tensile strength, fibers or yarns cannot even be processed. In addition, applications often impose restrictions on tensile strength, modulus, strain, or elastic recovery. Industrial fibers, for example, may need a minimum strength when subjected to imposed loads, or they may need minimal elongation when subject to imposed extensions. Fibers used in tire cords require both relatively high strength and high strain-to-break.

The requirements for torsional or bending properties may not seem so obvious at first, but in many applications they are more important than tensile properties. Consider, for example, applications in apparel. Clothes against the skin need to feel soft and comfortable. On the other hand, suit coat fabrics need to be rather stiff. The property that dominates fiber softness and stiffness is bending rigidity. The stiffness of a yarn is determined by the stiffness of the fibers, the number of fibers in the yarn, and the transfer of stresses among fibers in the yarn. The stiffness of a woven or knit fabric depends on that of the yarn and the construction of the fabric. The stiffness of a nonwoven fabric depends on that of its constituent fibers and on the details of the fabric construction and bonding. Other applications of the topics addressed in this chapter will become evident as the chapter progresses.

At the beginning of the chapter we confine our attention to small strains and deformations within the elastic limit. As in the case of tensile deformation, classical elastic mechanics may be used to calculate modulus and other important properties. Toward the end of the chapter we will deal with compressive strength and tenacity of fibers. As in the case of tensile deformation, linear elasticity theory is not applicable (e.g., $\sigma/\epsilon \neq$ constant) outside the Hookean region. For these and other reasons, it is difficult to predict strength. Modulus is a much better indicator of bulk fiber structure. Strength is discussed in detail in Chapter 9.

8.1 Basic Definitions

The following diagrams and equations define stresses and strains in tension and shear.

Tension

In the last chapter we presented the fundamental aspects of tensile deformation, shown in Figure 8.1:

$$\text{tensile stress,} \quad \sigma = F/A_0$$

$$\text{tensile strain,} \quad \epsilon = \Delta l/l_0$$

In the elastic region the tensile stress is directly proportional to the tensile strain. The elastic modulus, initial modulus, or Young's modulus is the proportionality constant:

$$\sigma = E\epsilon$$

The elastic modulus is a material property, which varies with direction in anisotropic materials. Poisson's ratio, ν, is a measure of the lateral contractile strain to the axial strain. If there is no volume change associated with deformation, Poisson's ratio is 1/2. Poisson's ratio is usually less than 1/2, indicating that volume increases in the Hookean elastic range. Hence, tensile deformation is often called dilatational or dilatant. Shear deformation, as we will see shortly, involves a change in shape with no change in volume.

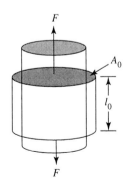

FIGURE 8.1   Schematic of tensile deformation of a cylinder.

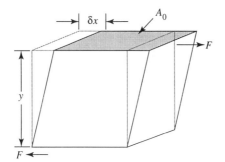

FIGURE 8.2 Schematic of shear deformation of a cube.

Shear

A schematic of shear deformation of a cube is shown in Figure 8.2.

$$\text{shear stress,} \qquad \tau = F/A_0$$

$$\text{shear strain,} \qquad \gamma = \delta x/y$$

In the elastic region, shear stress is directly proportional to shear strain. The shear modulus, G, is the proportionality constant:

$$\tau = G\gamma$$

The shear modulus is a material property, which varies with direction in anisotropic materials. Note that a tensile force is applied normal to the area used to determine tensile stress, whereas a shear force is parallel to the area used to determine the shear stress. We are now ready to begin our discussion of torsion and bending in fibers.

8.2 Torsion of Fibers

Torsion is shear with a different geometry than that just diagrammed, as shown in Figure 8.3. Since the displacement, δx, along F is $r\cos\theta = r\theta$ for small θ:

$$\text{shear strain,} \qquad \gamma = \theta(r/l)$$

$$\text{shear stress,} \qquad \tau = G\theta(r/l)$$

When torque is plotted on the ordinate and twist is plotted on the abscissa, a graph similar to the familiar tensile $\sigma-\epsilon$ curve is produced, as shown in Figure 8.4. Note that the shear stress and shear strain in the fiber or cylinder are proportional to the radial distance of the element of interest from the center of the fiber. That is, shear stress is not everywhere equal in the fiber, nor is the shear strain. As the torque is increased without limit, the fiber fails. The tenacity of various fibers in shear is shown in Table 8.1, along with the tensile tenacity of the same fibers. Twisting applies shear stresses and strains to the fiber. The maximum shear strain and stress are on the surface of the fiber.

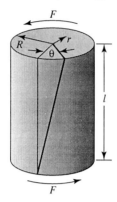

FIGURE 8.3   Shear deformation or torsion in a cylinder.

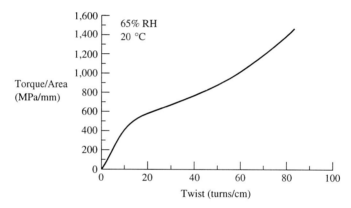

FIGURE 8.4   Torque–twist behavior of nylon. (W. E. Morton and F. Permanyer, *Journal of the Textile Institute* 49 (1949), T371.)

In practice the fibers fail when they reach a critical twist per unit length. The critical twist can be converted to a critical strain, a value that is independent of fiber diameter. The critical strain for failure of various fibers is reported in Table 8.2.

Let us return to the basic definitions of shear stress and shear strain. Another expression for differential shear stress in a round fiber or cylinder can be developed by considering the definition of shear stress:

$$\tau = df/da = df/2\pi r\,dr$$

Equating the two expressions for shear stress gives

$$G\theta(r/l) = df/2\pi r\,dr$$

Solving for the differential force,

$$df = 2\pi r^2 G\theta\,dr/l$$

TABLE 8.1   Shear and Tensile Tenacity of Fibers

Fiber	Shear Tenacity (g/d)		Tensile Tenacity (g/d)	
	65% RH	*Wet*	65% RH	*Wet*
Oriented cellulose	1.2	1.1	8.0	6.7
Cotton	1.0	0.85	2.7	2.4
Nylon	1.3	1.1	4.4	4.0
Viscose rayon	0.72	0.35	2.0	0.78
Acetate	0.66	0.57	1.3	0.88
Silk	1.3	1.0	3.5	2.8
Flax	0.91	0.84	2.9	3.2

Source: H. Bohringer and W. Schieber, cited by W. E. Morton and J. W. S. Hearle, *Physical Properties of Textile Fibers*, 3rd ed., Manchester, England: The Textile Institute, 1993.

TABLE 8.2   Critical Shear Strain for Failure by Twisting of Single Fibers

Fiber	Critical Shear Strain
Nylon—staple	1.66
industrial	1.15
PET—staple	1.66
industrial	1.07
Wool	0.84
Silk	0.81
Viscose rayon—staple	0.77
industrial	0.42
Cotton	0.71
Flax	0.48
Oxide glass	0.07

Source: P.-A. Koch, *Textil-Rondschau* 4 (1949), 199, and 6 (1951), 111.

This force produces a couple, rdf, or torque, which can be integrated over all radial shells to give the total torque:

$$\text{total torque} = \int rdf = \int_0^R 2\pi r^3 G(\theta/l)\,dr$$

which is just a sum over cylindrical shells. Solving,

$$\text{total torque} = \pi R^4 G\theta/2l$$

A cylindrical fiber will resist a torsional displacement, θ/l, by $\pi R^4 G/2$, which is called the torsional rigidity. Note that the resistance of the cylinder or round fiber contains both a material term, G, and a geometric term, $\pi R^4/2$. The geometric term contains the fiber radius raised to the fourth power. Reducing the fiber diameter by a factor of 2 reduces the torsional rigidity by a factor of 16. The geometrical effect is enormous.

A cylinder is generally used as the standard for assessing relative torsional rigidity. In ribbon-shaped objects, the torsional rigidity is proportional to ab^3

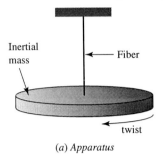

(a) *Apparatus*

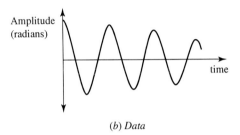

(b) *Data*

FIGURE 8.5   Torsion pendulum.

where a and b are the larger and smaller dimensions of the cross-section. Hence, ribbons are extremely flexible in twisting. Consequently, they are often used in conveyor-belt applications.

8.2.1 Measurement of Shear Modulus

A fiber's shear modulus can be determined from the torque–twist curve, much like Young's modulus is determined from a load–elongation curve; however, a much simpler and more elegant technique is available—the torsion pendulum. The apparatus is simple and easy to set up and use; in addition, accurate measurements can be taken, as shown in Figure 8.5. A fiber is subjected to a small shear strain imposed by a small initial twist. The period, T, and amplitude, A, of successive oscillations are measured, perhaps by noting each time the pendulum stops at its maxima and minima, as shown in Figure 8.5b. The period of the damped sinusoidal oscillation is determined using a stopwatch. The mass hanging from the fiber needs to be sufficiently small that tensile deformation does not occur, yet sufficiently large that the period of oscillation can be accurately measured. (A slightly more sophisticated experimental design allows elimination of the tensile strain caused by the weight and damping caused by air resistance on the inertial mass.) It can be shown that shear modulus depends on the period, the geometry, and the fiber properties. For a solid cylinder:

$$G = 2lM\omega^2/\pi R^4$$

TABLE 8.3 Torsional Properties of Fibers

Fiber (65% RH, 20 °C)	Shear Modulus (GPa)	Shape Factor (1.0 for round fibers)
Viscose rayon[a]	1.0	0.94
Silk (pie-shaped)[b]	2.4	0.84
Cotton (collapsed tube)[b]	2.2	0.71
Flax[b]	1.4	0.94
Wool (elliptical)[a]	1.3	0.99
Nylon 66[a]	0.4	1.0
PET[a]	0.8	1.0
Acrylic (bilobal)[a]	1.3	0.57
PP[a]	0.75	1.0
Spectra® 1000[c]	0.18	1.0
Kevlar® 49[c]	1.4	1.0
Kevlar® 149[c]	1.2	1.0
PBO[c]	0.97	1.0
Vectran® HS[c]	0.54	1.0
P100 Pitch-based C fiber[c]	4.7	1.0

Sources: [a]R. Meredith, *Proceedings Fifth International Congress on Rheology*, V1, University Tokyo, 1969, p. 43; [b]R. Meredith, *Journal of the Textile Institute* 45 (1954), T489; [c]V. Mehta and S. Kumar, *American Physical Society*, Seattle, March 1993.

where

G = shear modulus,

M = moment of inertia of the oscillating weight [$(1/2)mR_{disc}^2$ for cylindrical disc],

l = length of fiber,

ω = frequency of oscillation [$2\pi/T$], and

R = radius of cylinder or fiber.

The torsional properties of a number of fibers are provided in Table 8.3. Shape factors are used when assessing nonround fibers with equal cross-sectional area. The values for synthetic fibers are given as guidelines, since the molecular orientation and, hence, shear modulus are structure-dependent. The torsion pendulum is considered to be a dynamic oscillatory test in which information is learned regarding shear modulus. Other information can be learned using the torsion pendulum, and we will return to this technique in the next chapter. We will also investigate the tensile and bending equivalents of the torsion pendulum.

8.2.2 The Relationship Between Tensile and Shear Modulus

At this point you may ask how tensile modulus, E, and shear modulus, G, compare. The answer belies classical mechanics. For isotropic materials:

$$E/G = 2(1+\nu)$$

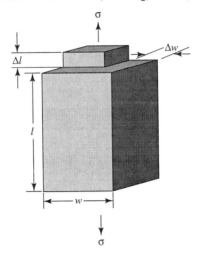

FIGURE 8.6 Dimensions required for the definition of Poisson's ratio.

where

$$\nu = \text{Poisson's ratio}$$

Recall that Poisson's ratio is defined as the ratio of lateral contractile strain to axial strain in a tensile test, as shown in Figure 8.6:

$$\nu = -(\Delta w/w)/(\Delta l/l)$$

where

    w = sample width,
    Δw = the change in width,
    l = sample length, and
    Δl = the change in length.

Like initial modulus, Poisson's ratio is measured only in the Hookean or linear elastic region. It is a measure of the change in volume or density of a material subject to a tensile deformation. Poisson's ratio for fluids is 0.5, since their volume is invariant with deformation. Polymers typically range from 0.3 to 0.4 and, hence, E/G for isotropic polymers is 2.5 to 3.0. For anisotropic polymers, the situation is much different. As molecular orientation increases in the fiber direction, E measured in the axial direction increases sharply, as shown in Figure 8.7. (E measured transversely decreases somewhat with molecular orientation.)

The structure of highly oriented fibers is fibrillar, as we saw in Chapter 3. We anticipate that tensile modulus will increase and shear modulus will decrease as the structure becomes increasingly fibrillar. A bundle of poorly bonded fibrils is more easily deformed by twisting than is a solid rod, since torsional resistance varies with R^4. Consequently, shear modulus decreases

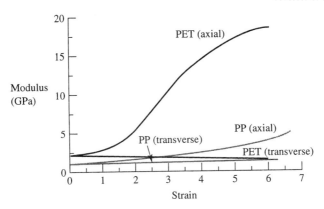

FIGURE 8.7 Effect of molecular orientation on tensile modulus. (D. W. Hadley, P. R. Pinnock, and I. M. Ward. *Journal of Material Science* 4 (1969), 152. Courtesy of Chapman and Hall.)

with increasing molecular orientation and E/G can become quite large, as shown in Table 8.4. E is chiefly determined by the extent of molecular orientation and G by the ability of the fibrils to deform uncoupled. Very high E/G fibers find use in ballistics as bulletproof vests. It is the tremendous area of new surface that develops when each fiber is split into its component fibrils, a process called fibrillation, that absorbs the ballistic energy. But we are getting ahead of ourselves, since fibrillation in ballistic applications is brought on in part by rapidly applied *compressive forces*. We will learn later in this chapter that the loose coupling of fibrils also gives these same fibers poor axial compressive strength.

TABLE 8.4 E/G of Various Fibers

Fiber	Tensile Modulus/Shear Modulus
Oxide glass[a]	2.0
Steel[a]	2.8
Cotton[a]	3.7
Flax[a]	19
Viscose rayon[a]	8.2
Acetate[a]	8.1
Wool[a]	3.2
Silk[a]	3.9
Nylon[a]	5.8
Kevlar® 49[b]	89
Kevlar® 149[b]	154
PBO[b]	330
Vectran® HS[b]	120

Sources: [a] W. E. Morton and J. W. S. Hearle, *Physical Properties of Textile Fibers*, 3rd ed., Manchester, England: The Textile Institute, 1993; [b] V. Mehta and S. Kumar, *American Physical Society*, Seattle, March 1993.

8.3 Bending

Woven and knit fabrics are made using yarns, which are aggregates of oriented fibers. Nonwoven fabrics are made directly using fibers. To bend a woven or knit fabric requires bending of the yarns. To bend a nonwoven fabric requires bending the fibers. In each case the fabric structure is also important. The resistance to bending is an important practical fabric characteristic. Bending stiffness relates to drape and comfort. In order to anticipate the bending properties of yarns and fabrics, we must first understand the bending properties of fibers.

Consider a rectangular beam subject to a bending stress, as shown in Figure 8.8. The longitudinal stress a distance y from the neutral surface is

$$d\sigma = E d\epsilon = E(y/\rho)$$

where

ρ is the radius of curvature of the beam, and
E is the modulus of the beam.

Since the width is b, the longitudinal force is

$$\sigma A = df = E(y/\rho)b\,dy$$

In pure bending the total longitudinal force is zero, since the compressive and tensile forces must balance:

$$(E/\rho)\int yb\,dy = 0$$

Each layer of force exerts a moment $y\,df$ about the horizontal (x) axis, tending to rotate the cross-section about the x axis:

$$M = (E/\rho)\int y^2 b\,dy = EI/\rho$$

where

$$I = \int y^2 b\,dy \equiv \text{second moment of area or moment of inertia}$$

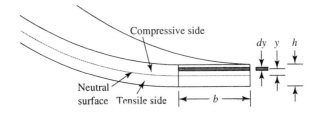

FIGURE 8.8   Bending of a rectangular beam.

TABLE 8.5   Flexural Rigidity of Various Fibers

Cross-Sectional Shape	Fiber	Shape Factor
Circle	round	1.00
Crenulated	viscose rayon	0.74*
Deeply crenulated	acetate	0.67*
Elliptical (almost round)	wool	0.80*
Pie-shaped	silk	0.59*
Square	synthetic	1.05
Equilateral triangle	trilobal	1.21
Rectangle $b \times 4b$	synthetic	0.26

Source: *Modified from D. Finlayson, *Journal of the Textile Institute* 37 (1946), P168.

Here, for a rectangular beam, b = constant and, consequently,

$$I = b \int_{-h/2}^{h/2} y^2 \, dy = bh^3/12$$

Thus, the flexural rigidity, EI, of a rectangular beam or slab is proportional to h^3, or the third power of the slab thickness. We will use this expression when assessing the bending stiffness of some bonded nonwoven fabrics. Indeed, a common complaint about most nonwoven materials is that they are "boardy." Fabrics seem stiff and boardy when fibers or yarns cannot slide by one another in response to shear stresses.

For a cylindrical fiber, we proceed with a calculation of the moment of inertia about both x and y axes:

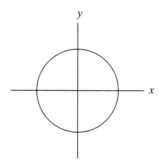

$$I = \tfrac{1}{2}[I_x + I_y]$$

$$= \tfrac{1}{2}\left[\int_A x^2 \, dx\,dy + \int_A y^2 \, dx\,dy\right]$$

But $x^2 + y^2 = r^2$ and the differential area is $dx\,dy = 2\pi r \, dr$. Therefore,

$$I = \tfrac{1}{2}\int_0^R r^2(2\pi r)\,dr = \pi R^4/4$$

Thus, the bending rigidity of a cylindrical fiber can be expressed as

$$EI = E\pi R^4/4$$

Like the torsional rigidity, the bending rigidity contains a material constant, E (the tensile modulus), and a geometrical constant, $\pi R^4/4$, which contains the diameter to the fourth power, or the cross-sectional area squared. As in torsion, shape factor is used to assess the bending rigidity of nonround fibers. Simply multiply the shape factor by the bending stiffness of a round fiber with equal cross-sectional area to adjust for fiber cross-sectional shape. Shape factors for various fibers and geometries are given in Table 8.5. Keep in mind that synthetic fibers can be extruded to provide a range of cross-sectional shapes and sizes. When a flexible synthetic fiber is required, first choose a material with a low modulus. Then select a fiber with a low denier and focus on fibers with ribbonlike cross-sections. High twist levels create a structure that is difficult to bend: A highly twisted yarn acts like a rod, since the fibers are not free to slip at the interface. Let me remind you once again that bending stiffness is proportional to diameter to the fourth power. The popularity of multistrand copper wire is due in large part to its ability to bend more easily than solid core wires.

8.4 Compression

A compressive stress is simply a negative tensile stress, and a material's response is a negative strain. The initial compressive modulus is generally the same as the initial tensile modulus; however, the compressive strength of a material often differs significantly from its tensile strength. Ceramics, such as aluminum oxide fibers, have much higher compressive than tensile strength. Organic fibers, on the other hand, typically have much lower compressive than tensile strength. A typical stress–strain curve for a highly oriented organic fiber is shown in Figure 8.9. When a fiber is subjected to axial compression, it will buckle; however, if the fiber is supported on its sides, then buckling may be avoided. Failure occurs by other mechanisms. We will first treat buckling à la Euler, and then treat compressive failure of fibers used in reinforcement.

8.4.1 Buckling of a Column

The axial compressive properties of fibers may be important in textile applications. An example of an instance in apparel use when buckling is desirable is the folding of a fabric, say the sleeve of a sweater when the arm is bent. When they are unconstrained, long spans of fiber subject to axial compression, as shown in Figure 8.10, typically bend in an unstable fashion and buckle. The resistance to buckling is given by Euler's equation:

$$P_{\text{critical}} = \pi^2 EI/L^2$$

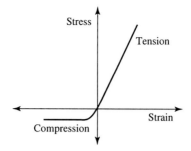

FIGURE 8.9   Tensile and compressive behavior of highly oriented fiber.

FIGURE 8.10   Axial compressive loading of a cylinder.

where

P_{critical} = critical load above which buckling occurs,

EI = flexural rigidity, and

L = column height.

Hence, buckling can be avoided when the span of the fiber is small compared with the smallest cross-sectional dimension of the fiber. Buckling can be largely avoided in dense composite structures, in which the matrix provides lateral constraint to the fibers. Similarly, fiber buckling can be minimized in yarns by imparting a high level of twist. When a yarn behaves as a coherent entity, such as a yarn with a high level of twist, then the yarn behaves essentially as a rod rather than as a collection of independent fibers. Hence, a highly twisted yarn, such as a worsted yarn as contrasted to a woolen yarn, has a high resistance to buckling. In the absence of buckling, our concern shifts to the fiber's compressive strength.

8.4.2 Axial Compressive Properties

It has been known for about four decades that oriented fibers fail in compression at surprisingly low strain, displaying bands of shear deformation, as shown in Figure 8.11. When a fiber is bent to a small radius of curvature, the inside of the loop experiences compression and the outside of the loop experiences tension. The advent of Kevlar® and other high-performance organic fibers brought the realization that compressive properties of fibers are important in dense composite structures. Consider, for example, the technology required to enable a submarine to explore vast depths of the oceans. Such structures need to be composites, probably on the order of a meter thick, and capable of withstanding high compressive stresses.

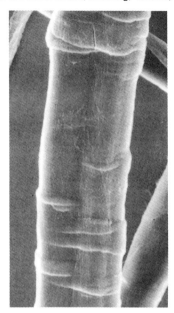

Flax Fiber
(Courtesy of The Institute of Paper Science and Technology.)

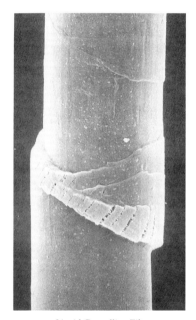

Liquid Crystalline Fiber
(V. Mehta and S. Kumar, *American Physical Society*, Seattle, March 1993.)

FIGURE 8.11 Scanning electron micrographs of kink bands in oriented polymers.

TABLE 8.6   Axial Compressive Strengths of Reinforcement Fibers

Fiber	Tensile Strength (GPa)	Compressive Strength (GPa)
Spectra® 1000	3.0	0.2
Kevlar® 149	3.4	0.4
PBO	5.0	0.3
Carbon, T-300 PAN-based	3.2	2.8
Carbon, P-100 pitch-based	2.2	0.5
Oxide s-glass	4.5	1.1
Aluminum oxide	1.7	6.9

Source: S. Kumar, *Indian Journal of Fiber and Textile Research* 16 (1991), 52.

One role the matrix serves in a composite structure is to provide lateral support to the fibers, which prevents them from buckling. Therefore, when a composite is subjected to compressive forces along the fiber axis, the fiber itself needs to have high compressive strength. Unfortunately, organic fibers require high molecular orientation to elevate tensile strength and modulus, and hence, organic fibers typically have poor axial compressive properties. As mentioned, the compressive modulus is the same as the tensile modulus, since the atomic bonds are stressed similarly. The problem is that the compressive strength or tenacity is low, as shown in Table 8.6. The compressive strength is poor in high-modulus organic fibers because the fibrils or molecules themselves have little lateral interaction, allowing each fibril or molecule to buckle at low strain. The process has been termed microbuckling or shear banding. A notable exception to the rule that all high-strength and high-modulus fibers fail in compression by shear banding is spider drag line silk. This fiber has a tenacity of about 18 g/d, but the fiber does not fail in compression when subjected to bending strains of 20 percent or more. Other high-strength synthetic organic fibers fail (to carry increasing loads) at less than 1 percent compressive strain.

One potentially useful application of shear banding is in the area of elastic recovery of films. Aluminum foil is useful as a food wrap because it has the property of dead bend, which is virtually zero elastic recovery from a bending strain. Achieving dead bend in polymers is nontrivial, but recovery from shear banding is indeed very low.

8.4.3 Transverse Compressive Properties

The transverse compressive properties of most interest are modulus and resilience. The transverse modulus is important in modeling the mechanical behavior of fibers and composites made of fibers. The properties of the fibers influence the resiliency of a thick fabric. Resiliency from lateral compressive

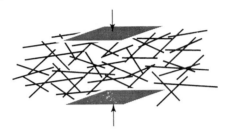

FIGURE 8.12   Compressive deformation of a mat.

forces is especially important in nonwoven structures used for insulation and other high loft nonwoven applications, as illustrated in Figure 8.12. The compressive properties of such an assembly depend largely on the fiber bending, which has both material and geometrical aspects, but also on transverse fiber compression.

Information on the transverse modulus is available in the published literature. Variation in the transverse modulus with molecular orientation is shown in Figure 8.3, and data on transverse modulus are compiled in Table 8.7. Like the relationship between the axial tensile and compressive modulus, the transverse compressive modulus is equal to the transverse tensile modulus. High-modulus organic fibers have low transverse moduli. Oxide glass is isotropic. The lateral compressive strength of highly oriented fibers is also poor. Fibers with molecules wrapped helically about the fiber, such as milkweed and eucalyptus, have good lateral compressive strength. Bulletproof fabrics were discussed earlier in the text. Projectiles impact fibers laterally and cause the strong fibers to fibrillate rather than to fracture normal to the axis. The new surface has an energy associated with it, γ_{S-V}. Since the fibrils are small in diameter, much new surface area, A, is formed, absorbing energy associated with the ballistic impact. The fibrils themselves probably have a strength com-

TABLE 8.7   Transverse Modulus of Fibers

Fiber	Axial Modulus (GPa)	Transverse Modulus (GPa)
Nylon	4.3	1.2
PET	16	1.2
PP	5	1.0
Carbon fiber	168	6.8
Carbon fiber	379	3.1
Kevlar® 149	179	2.5
Spectra® 1000	89	1.2
Oxide glass	77	68

Source: S. Kawabata, "Measurements of transverse mechanical properties of high-performance fibers," *Journal of the Textile Institute* 81, 4 (1991), 432–447.

parable with that of the fiber, although each location of the vest where an impact occurred needs to be replaced.

8.5 Summary

Fibers are often used in applications that demand properties other than tensile properties, such as torsion, bending, or compression. The resistance to torsion and bending of a solid round fiber increases with the cross-sectional area squared. This explains the historic value placed on fine fibers in apparel applications. The highest-quality wool fibers are the smallest diameter fibers. Woolen yarns with fine fibers can lead to very soft fabrics. Flocked fabrics are ones that have individual fibers standing up on the surface. These materials are also quite soft. Meltblown polypropylene webs can have very small fibers, and fabrics made from the small fibers are also extremely soft.

The axial compressive properties of fibers are also important, most certainly in apparel and composite applications. The resistance to buckling of a round fiber is proportional to its diameter raised to the fourth power and inversely proportional to the span squared. The compressive strength of a fiber decreases with increasing molecular orientation. At very high molecular orientation, the compressive strength may be as low as 5 percent of the tensile strength. On the other hand, tensile and compressive moduli are the same in axial testing. They are also the same in transverse testing, but generally lower than in the axial direction. The ratio of tensile to shear modulus, $E/G = 2(1 + \nu) = 2.4$ to 3.0 for an isotropic material such as an oxide glass or epoxy, but can exceed 100 for highly oriented fibers. The low shear modulus and low compressive strength of highly oriented fibers is due to the weak bonds between fibrils, which allow the fibrils to act independently. A bundle of fibrils is easier to buckle in compression and twist in torsion than is a solid rod.

Of all the fibers with a solid cross-section, round fibers offer very high resistance to shear or torsion, bending, and buckling. Cotton fibers, being ribbon-shaped in cross-section, are relatively easy to bend or twist.

References

A. H. Cottrell. *Mechanical Properties of Matter*. New York: J. Wiley, 1964.

S. Kumar and S. Lee, ed. *International Encyclopedia of Composites*, V4. New York: VCH, 1991.

F. McClintock and A. S. Argon. *Mechanical Behavior of Materials*. New York: J. Wiley, 1967.

R. Moreton, W. Watt and B. V. Perov, eds. *Strong Fibers*, V1, *Handbook of Composites*. New York: Elsevier, 1985.

W. E. Morton and J. W. S. Hearle. *Physical Properties of Textile Fibers*, 3rd ed. Manchester, England: The Textile Institute, 1993.

S. B. Warner, "Dead Bend Plastic Food Wrap," US Patent No. 4,882,230, 1989.

Problems

(1) Using values for bending modulus and fiber dimensions, calculate the specific bending rigidity of a wool fiber. State which type of wool you have considered.

(2) Suppose you have two (round) fibers that are identical except that fiber 2 has a diameter three times that of fiber 1. How does their resistance to tensile bending and torsion deformation differ (quantitatively)?

(3) Compare the bending rigidity of a single monofilament with cross-sectional area A with that of a bundle of 100 fibers each with area $A/100$. Assume friction is negligible and the small fibers act independently.

(4) Calculate the ratio of bending resistance of a solid round fiber with a diameter of 20 μm with that of a hollow round fiber with the same denier. The hole in the center of the fiber (shown) is 8 μm in diameter.

(5) Qualitatively compare the resistance to buckling, torsion, and bending of a woolen versus worsted yarn. Describe a good application for a worsted yarn.

(6) Consider an optical monofilament of isotropic poly(methylmethacrylale), PMMA, that is 1.0000 mm in diameter. A 3.000 cm length is elastically extended to 3.010 cm, and the simultaneous diameter reduction is 1 μm. Calculate Poisson's ratio.

(7) Consider the use of both round polypropylene microfibers, each about 5 μm in diameter, and PET staple fibers, each about 25 μm in diameter, to make a lofty nonwoven bat. (This is similar to the composition of Thinsulate®, a 3M product used for thermal insulation.) How much more resistance to lateral compression does the PET provide compared with the PP?

(8) Silk and wool are considered for use in woven fabrics that will experience torsion. Compare the two fibers with respect to **(a)** inherent material properties and **(b)** effects related to fiber shape.

(9) Poisson's ratio can be difficult to measure on fibers. Can E/G be used to assess Poisson's ratio?

(10) Given that Poisson's ratio for a material is 0.38, how much volume change occurs with 4% strain in the Hookean elastic region?

(11) Kink bands may reduce the tensile strength of fibers significantly, but not as much as might be anticipated, i.e., generally less then 25% strength loss. If, however, you are using the fibers in an environment that attacks the fiber and you were relying on slow diffusion as a means of protection, such as in a tire, what might occur in fibers with kink bands?

(12) A sample of nylon monofilament (fishing line) is tested using a torsion pendulum. The fiber radius is 127 μm, the test length is 17.8 cm, and the moment of inertia of the disc is 1.29×10^{-6} kg-m^2. The period of oscillation is determined to be 6.54 sec. Calculate the shear modulus. Compare the shear modulus with the tensile modulus, 2.5 GPa.

(13) Calculate the percent volume change associated with twisting a 30 denier PET fiber to a strain of 0.5%.

(14) Show why the only material constant involved in bending a fiber is the tensile modulus, yet the material in the fiber may be in tension, compression, or unstressed.

9

Variability and Control

In previous chapters we discussed the fact that cotton and other natural fibers are highly variable in length, denier, cross-section, density, tenacity, modulus, and elongation-to-fail, as well as in other physical properties. We also noted that synthetic fibers are typically variable in tenacity, denier, modulus, staple length, and the like, largely because fiber processors do not go to the added and unnecessary expense to control these properties. In this chapter we will learn about the effects of variability in natural and synthetic fibers. In the latter part of the chapter we will explore techniques to monitor and control fiber variability. In industrial fibers, where high-strength or high-modulus yarns are required, minimization of variability is imperative. Even in textile fiber operations, such as forming, knitting, or weaving of fine threads, it is important that variability be contained.

We will speak of both fibers and yarns in this chapter. Yarns are made of fibers, either continuous filaments or staple fibers. Staple fibers may be natural, synthetic, or blends. In natural fibers, virtually every property is variable from point to point, fiber to fiber, and lot to lot. In synthetic fibers, the polymer chemistry, staple fiber length, modulus, tenacity, denier, and so on are usually less variable than in natural fibers. Texturing, which imparts crimp to a synthetic fiber, may add to the variability.

Applications for single fibers are few. Fibers are often combined into yarns to form a useful product. A fine denier yarn requires fine denier fibers. Otherwise, the uniformity is hopelessly poor. A yarn with low variability does not necessarily require fibers with low variability. Variability depends on the scale of measurement or investigation. We will measure the scale of variability using frequency only briefly toward the end of the chapter.

9.1 Intrinsic and Extrinsic Properties

One of the most important distinctions among properties is whether they are intrinsic or extrinsic. An intrinsic property is one that depends on the bulk morphology of a material. Modulus and density are good examples. If a small part (say 0.1 mm long) of a fiber sample (say 10 cm long) has an unusually low modulus or density, for example, it will impact the modulus or density of the entire sample only minimally. If a small part of a sample has a low strength, on the other hand, then the strength of the entire sample is affected. Strength is an extrinsic property. It is determined by only a small part of the sample.

You are probably familiar with the use of the terms *intrinsic* and *extrinsic* in semiconductor physics. An intrinsic semiconductor is a pure material that has a bandgap of a few tenths of an electron volt. An extrinsic semiconductor is a material that needs to be doped with a few parts per million of another material in order to give it an effective bandgap of a few tenths of an electron volt. The terminology is consistent. Similar terminology is used in thermodynamics for intensive and extensive properties.

Let us examine one of the most important extrinsic properties of materials: strength, or tenacity. We cannot say as much about strength as we can about most intrinsic properties, nor can we learn as much about bulk structure by examining strength as we can by examining intrinsic properties such as modulus. Tenacity, however, is a very important engineering property of fibers. It is imperative that we learn the guiding principles of strength and how to "predict" strength.

9.2 Fundamentals of Strength

The theoretical strength of a material can be calculated rather simply using some basic arguments and approximations. Consider, for example, a crystal of material. The energy required to separate atoms is the sum of the energies holding the atoms together. When the atoms have slipped one lattice spacing, they are again in their lowest energy state. We assume the shape of the stress–displacement curve is sinusoidal, and failure occurs at a critical stress, σ_c, as shown in Figure 9.1. We can immediately write an expression involving critical stress:

$$\sigma = \sigma_c \sin(2\pi x/\lambda)$$

which for small values of the displacement is approximately

$$\sigma = \sigma_c(2\pi x/\lambda)$$

Continuing with the assumption that displacements are small, Hooke's law is obeyed:

$$\sigma = E\epsilon = Ex/a_o$$

(a) *Stepwise slippage of one row of atoms over another*

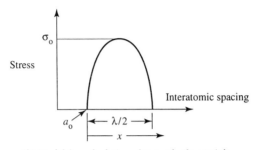

(b) *Model for calculation of strength of materials*

FIGURE 9.1   Theoretical strength of materials.

and then

$$\sigma_c = \lambda E / 2\pi a_o$$

If all the work done in causing fracture goes into forming new surface area with energy γ, then each time a bond is broken, two new surfaces are formed:

$$2\gamma = \int_0^{\lambda/2} \sigma_c \sin(2\pi x / \lambda) \, dx = \lambda \sigma_c / \pi$$

so that

$$\sigma_c = (E\gamma / a_o)^{1/2}$$

When we perform the calculation indicated, we arrive at the conclusion that the theoretical strength of a material is roughly $E/10$, or 10 percent of Young's modulus. When we go into the laboratory and measure the strength of materials, however, we find that most materials have a strength of $E/1000$, or 0.1 percent of Young's modulus. Take oxide or window glass as an example. Using a testing machine, we can determine that the modulus is about 70 GPa and the strength is about 70 MPa. Why are materials so much weaker than their theoretical strength? The reason is that defects are present. Since strength is an extrinsic property, a defect can have a significant effect on strength. The severity of the worst defects in a test volume or fiber length will determine the strength of that sample of fiber.

The defects in most materials can be envisioned as cracks. They may not really be cracks, but cracklike models have been very successful at predicting the strength of materials. Consider a crack in a material oriented normal to the test direction, as shown in Figure 9.2. When the material is loaded as shown, the stress is magnified at the tip of the crack by what is called the

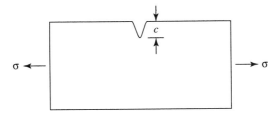

FIGURE 9.2 Schematic of edge flaw in a material.

stress intensification factor. For a circular hole in a flat plate, for example, the stress is magnified three times at the crack tip. For a sharp crack the stress intensification factor is much greater than three.

We can calculate the effect of the crack on the strength of the material quite easily if we make a few assumptions. These assumptions were first made by Griffith in a classic 1920 paper when he was dealing with brittle materials like oxide glass (*Philosophical Transactions of the Royal Society* A221 (1920), 163). We will make the same assumptions to show the behavior of materials, but recall that few polymers are ideal brittle materials. The chief assumption we make is that all the energy put into the material by the stress, $\sigma d\epsilon$, goes into forming new surfaces that develop when the crack propagates, γdA. No energy is converted to heat by internal friction, such as atoms sliding past one another. Putting this in differential terms for an elliptical crack with major axis $2c$ in a thin plate:

$$d[\pi c^2 \sigma^2 / E]/dc = d[4\gamma c]/dc$$

Equating the terms in the brackets and solving for σ gives

$$\sigma = (2E\gamma/\pi c)^{1/2}$$

From this expression, appropriately called the Griffith equation, we can readily observe that as crack length increases, strength decreases. Hence, in two fibers that are nominally identical, the strengths will differ according to the length of the microscopic cracks present. It can be extremely difficult to eliminate or tightly control crack size during manufacture of fibers. Defects can be introduced in polymerization, extrusion, drawing, or finishing in synthetic fibers, or by growth patterns or finishing in natural fibers. Many strength-limiting defects are surface cracks.

The preceding equation does not work in detail for organic fibers, since most polymer fibers show a certain amount of ductility. Organic fibers that contain oriented molecules are not as sensitive to flaws as are other materials, but the Griffith equation does highlight the basic behavior of highly oriented fibers. In conventional flexible chain semicrystalline fibers, the folds at crystal surfaces are generally regarded as the strength-limiting defects. So long as conventional fiber processing techniques are used, fiber tenacity seems to be limited by these sorts of defects to about 10 g/d. Fiber tenacity generally increases with molecular orientation, but not without limit. Eventually, still other defects are introduced, such as precracks or crazes.

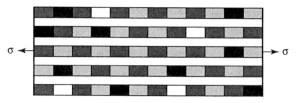

σ ← → σ

FIGURE 9.3 Model of five fibers tested in tension as an untwisted yarn with each fiber modeled as a chain of links.

We also note that the smallest crack that can exist in a material is one with atomic dimensions. If a crack with a length of a few Angstroms is substituted into the preceding expression, we calculate the strength is about $E/10$, just what we found using another technique. An oxide glass fiber that is manufactured under meticulous conditions of cleanliness may be as strong as 5.6 GPa, which is $E/12$!

9.2.1 The Weak Link Concept

We realize that a yarn is a collection of fibers. We need to know how fiber strength influences yarn strength. Peirce addressed this problem in a classic paper published in 1928 (F. T. Peirce, *Journal of the Textile Institute* 17 (1926)). Peirce was a statistical mathematician and he approached the problem statistically. He called his theorem the "weakest link theorem," which states that a yarn is like a chain. A chain will break at its weakest link. Consider the strength of pieces of an ideal yarn, a yarn in which all the pieces or links have the same strength, as shown in Figure 9.3. Consider now a real yarn, in which a small fraction of the pieces are weaker than the other pieces. The strength of the ideal yarn does not depend on test length; however, the strength of the real yarn does. If just one link is tested, then the pieces of the real yarn are the strength of the link—either weak or perhaps strong. At a long test length, though, the weak link has a high chance of being present, and the yarn will be as weak as the weakest link. Yarn strength depends on the coefficient of variation of strength, CV = standard deviation/mean. Peirce expressed his results using test or gauge length studies:

$$\sigma_{nl}/\sigma_l = 1 - 4.2(1 - n^{-1/5})CV$$

where

σ = strength, or tenacity,

n = ratio of the test length, large gauge/short gauge, and

4.2 and −1/5 are empirical constants.

The equation states that strength decreases with increasing test length when CV > 0, i.e., if there is variability. The larger the variability, the more the yarn weakens with test length. The fundamental reason that the strength decreases with test length is that the probability of finding a serious defect increases with sample size.

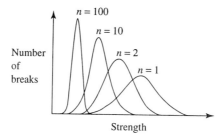

FIGURE 9.4   Effect of test length on yarn strength distribution.

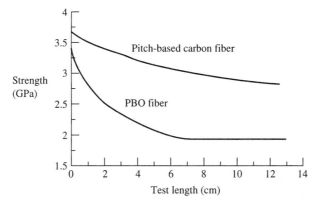

FIGURE 9.5   Effect of test length on strength of PBO and carbon fiber. PBO data: V. Mehta and S. Kumar, *American Physical Society*, Seattle, March 1993. Carbon data: R. Moreton in W. Watt and B. V. Perov, eds., *Strong Fibers*, V1, *Handbook of Composites*, New York: Elsevier, 1985.

Consider a yarn consisting of fibers that are characterized by a normal distribution in strength. The yarn is tested at various test lengths, from a length of unity to a length 100 times that. The calculated frequency of breaks is plotted as a function of the strength in Figure 9.4. The yarn is weaker at higher test lengths because of the increased probability of defects in the test section. The frequency distribution narrows with test length because at high test length the probability of a defect appearing in the test length approaches one, suggesting that a defect is almost always present. At small test lengths some samples are highly defective and others are not. Peirce's equation gives a quantitative expression for the strength of the yarns at a large test length, relative to that at a smaller test length, given the standard deviation in fiber strength at small test length.

The work of Peirce is empirical and was conducted on cotton fiber yarns. His results are general, however, and data support the expansion of his theory to most fibers. For example, the data in Figure 9.5 show that the tensile strength of PBO and carbon fiber is highly sensitive to test length.

A corollary to Peirce's work is that fiber strength generally decreases with fiber diameter. At a given test length small diameter fibers have less chance of

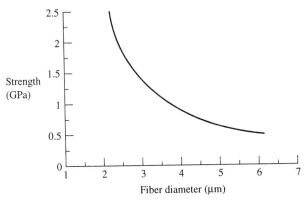

FIGURE 9.6 Effect of fiber diameter on strength of Saffil® alumina fiber. (J. D. Birchall, J. A. Bradbury, and J. Dinwoodie in W. Watt and B. V. Perov, eds., *Strong Fibers*, V1, *Handbook of Composites*, New York: Elsevier, 1985.)

a major defect existing in the test section. Experimental data for oxide glass, carbon, and other fibers confirm the prediction. The data shown in Figure 9.6 are for alumina fiber.

Let us point out the consequences of Peirce's analysis and for the moment ignore the quantitative aspects. Natural fibers can have numerous defects that arise during growth and others that may develop during processing. The CV in strength among fibers may be very large. Hence, although mean fiber strength may be reasonable, the yarn strength may be poor. Cellulose indeed can have very high strength. Wood pulp fibers (Spruce) have single-filament tenacity as high as 12 g/d (D. H. Page et al., *Int. Paper Phys. Conf.*, PPRIC, Mont Gabriel 20–23 Sept. 1971, and C. Y. Kim et al., *Journal of the Applied Polymer Science* 19 (1975), 1549), yet cotton yarns are typically only 3 g/d. Synthetic textile fiber processing is generally not one of good control. Hence, the CV in strength is high, perhaps 20 percent or more. Synthetic fibers may have high individual strengths, such as 9 g/d for PET; however, the yarn strength is lower. Textile yarns generally do not need to be stronger than about 4 g/d. Industrial fibers, such as those used in composites, are generally manufactured under conditions of high control. Hence, the CV in strength is low. In some fibers, such as Kevlar® and Spectra®, data suggest that the defects that limit strength may be as small as individual chain ends (Y. Termonia, P. Meakin, and P. Smith, *Macromolecules* 18 (1985), 2246).

Peirce's analysis provides a good understanding of the effect of flaws on yarn strength. There may be other sources of defects, however, that Peirce did not consider. Consider, for example, systematic (periodic) variations in a property, such as denier. Such a defect as this, which is not random, but fluctuating, can develop on fiber extrusion and take-up. Peirce's analysis, as well as others based on random flaws, will not fit the data (W. Knoff, *Journal of Materials Science* 22 (1987), 1024–1030, and 28 (1993), 931–941). Other theories have been developed for these sorts of problems, but they are well beyond the scope of this text.

9.2.2 Weibull Analysis

In 1939 Weibull introduced a statistical theory that is often used to handle failure in materials that are flaw sensitive (W. Weibull, *Vetensk. Akad. Proc.*, No. 153 (1939), 151). Like other theories, Weibull's treatment makes assumptions about the probability of occurrence of critical flaws in a specimen volume or surface area. Weibull assumed the risk of rupture, R, is proportional to a function, f, of the volume, V, and stress, σ, on the body:

$$R = \int f(\sigma)dV$$

An explicit function for $f(\sigma)$ is obtained by integrating over the volume subjected to the stress:

$$f(\sigma) = (\sigma/\sigma_o)m$$

where

σ_o = characteristic strength, which depends on the distribution function, and

m = constant related to the homogeneity of the material.

As $m \to 0$, the maximum of inhomogeneity, $f(\sigma) \to 1$. Hence, the risk of rupture is independent of applied stress. At the other limit of m, $m \to \infty$, $f(\sigma) \to 0$ for all values of σ less than σ_o. Fracture always occurs when $\sigma = \sigma_o$. An average value for strength is given by

$$(\sigma)_{R=1/2} = \sigma_o(1/2)^{1/m}$$

The ratio of $(\sigma)_{R=1/2} : \sigma_o$ is a measure of the variability of strength. Weibull's theory correctly predicts that strength decreases with increasing test or sample size. It is currently used extensively to analyze data on fibers for composite reinforcement and in ionic solids.

9.2.3 Effect of Twist on Strength

The effect of yarn twist has been discussed in two places in this text, and the results are summarized schematically in Figure 9.7. In Chapter 3 the effect of staple fiber length and twist on strength was presented, and in Chapter 7 the effect of helix angle on strength was derived. Twist gives staple yarns strength by allowing transfer of loads among fibers through friction. Following Peirce, twist, then, can increase strength by inactivating defects (fiber–fiber stippage). Too much twist, however, will reduce strength by the second power of the cosine of the helix angle. Yarn twist reduces the effective test length of the fibers, essentially dividing a test length into a number of smaller ones.

Whenever a property such as strength in twisted staple yarns depends on two functions, one that increases with a variable and the other that decreases with the variable, then the resultant property shows a maximum. Here we see

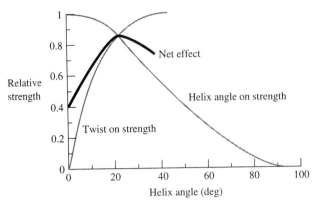

FIGURE 9.7 Effect of twist on modulus and strength of yarns.

that staple yarn strength is a maximum at a helix angle of about 20°. The stiffness of the yarn, however, also increases with twist.

9.3 Fundamentals of Modulus

As discussed earlier, modulus is an intrinsic property. It is not especially sensitive to the presence of defects. If a defect reduces the effective cross-sectional area by, say, x percent, then the modulus in that area will decrease x percent also, all other factors being equal. Table 9.1 shows the within-sample variability in cross-sectional area for a number of natural and synthetic fibers. Wool is characterized by a very high variability in denier, nearly 40 percent. Synthetic staple textile fibers may show denier variability of as much as about 20 percent, but 10 to 15 percent is typical. Industrial fibers have lower denier CVs, perhaps 10 percent or less.

Being an intrinsic property, modulus is a good indicator of the bulk structure of a fiber. In fact, modulus is determined by the chemical structure of the polymer and the molecular orientation of the fiber. The more linear and stronger the covalent bonds, the stiffer the molecule. Table 9.2 give the theoretical modulus of various polymers. The values cited are for molecules perfectly aligned along the fiber axis. The increase in axial modulus with molecular alignment occurs at the expense of transverse modulus; however, fiber scientists are not usually concerned with transverse modulus. The effect of net molecular orientation on fiber modulus is shown in Figure 9.8. In general, modulus increases with draw ratio or strain.

Normal to the axis of an oriented fiber, the strength and modulus are inferior to those along the fiber axis. High strength and modulus in polymers goes hand in hand with high anisotropy. We will learn a few techniques to characterize fiber anisotropy in Chapter 12.

The effect of a helical molecular conformation on fiber modulus is similar to the effect of twist on yarn modulus. The more highly oriented are the

TABLE 9.1   Variability in Denier of Various Fibers

	Coefficient of Variation (%)	
Fiber	*Denier*	*Tenacity*
Cotton	24	43
Silk	17	20
Wool	21	28
Rayon	12	17
Nylon (industrial fiber)	9	7

Source: R. Meredith, *Journal of the Textile Institute* 36 (1945), T107.

TABLE 9.2   Theoretical Modulus of Polymers

Polymer	Modulus (GPa)
Graphite	1500
PBO	730
PPTA	250
Polyethylene	360

Source: S. G. Wierschke in W. W. Adams, R. K. Elby, and D. E. McLemore, eds., *The Materials Science and Engineering of Rigid-Rod Polymers*, Materials Research Society Symposium Proceedings 134, Pittsburgh, PA, 1989.

molecules along the fiber axis, the more chains bear the load, and the stiffer the fiber. (These same arguments apply to strength, but the presence of defects controls ultimate stress.) In the specific case of a yarn or molecule with a constant helix angle, θ, the modulus decreases with $\cos^4 \theta$.

Modulus and other intrinsic properties can vary in fibers and yarns along the fiber length, and it may be important to monitor and control them. Keeping apprised of molecular orientation is one way to indirectly monitor modulus. Similarly, continuously measuring the load required to maintain a fixed

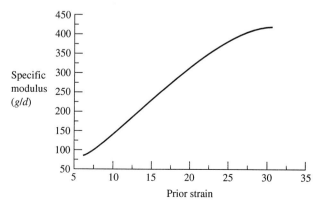

FIGURE 9.8   Effect of molecular orientation on modulus of polyethylene.

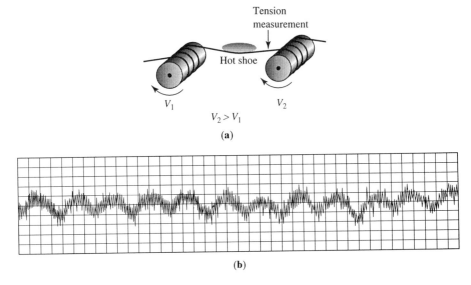

FIGURE 9.9 (a) Device for assessing load at specified elongation in filament yarn. (b) The load required to maintain fixed elongation (25%) of continuous filament polyester spun at 4850 m/min is shown on the ordinate. The average force was 154 g and the range was 22 g. The abscissa is time and the results shown correspond to 6 m of yarn.

elongation (25 percent here) gives a single-point determination of a modulus, as shown in Figure 9.9b. To obtain these data, PET filament yarn was drawn over a hot shoe between two rolls turning at different speeds, as shown in Figure 9.9a. This leads us directly into the following section, in which the various ways that a fiber or yarn can be analyzed to provide a measure of its uniformity are examined.

9.4 Monitoring and Control

It is a straightforward matter to measure properties of fibers or yarns after they are prepared: Compute means, standard deviations, and CVs and compare the values to those required. This technique is effective for tenacity, specific modulus, diameter or denier, and so on. Random sampling and data analysis should be conducted as suggested in Chapter 4 or in a statistics text. One of the chief shortcomings with this technique, however, is that when the production process is not operating as desired, the material may be substandard. Since the material is not tested until a significant quantity is produced, a large volume of waste or substandard material may be generated. Clearly it is advantageous to conduct on-line monitoring and testing, especially tests that can give insight into problems that may develop. Feedback loops may be deployed so that corrective action may be taken automatically.

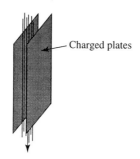
Charged plates

FIGURE 9.10   Principle of an evenness testing machine.

In order to take advantage of on-line testing, it is imperative that the processing system have sufficient flexibility that control is possible. When extruding 2000 fibers through a single spinneret, for example, it is not possible to control the properties of each fiber independently. Hence, control is poor and monitoring can be used only for limited purposes. This system has inherent position-to-position variability; however, the variability with time can be a different matter.

A number of techniques may be used to ascertain the characteristics or quality of a fiber or yarn as it is being produced or modified. Most of them are electro-optical. Some of the techniques I offer apply only to synthetic fibers during spinning or post-processing. Others apply to yarns, either natural, synthetic, or blends. Often, on-line measurements need to be nonintrusive, requiring contact-free devices. Optical and electrical devices suit this requirement, but when mechanical properties are needed, contact devices may be necessary.

9.4.1 Temperature- and Pressure-Sensing Devices

Monitoring may in many instances be achieved by sensing pressures or temperatures at various critical positions of processing. In fiber extrusion, for example, the spin pack temperature and pressure are critical to the formation of uniform filaments. Monitoring and control of this sort are commonly used in chemical and physical processes.

9.4.2 Mass-Sensing Devices

Techniques to measure denier or mass of yarn passing a certain point with time are numerous. The outward pressure exerted by a yarn passing through a funnel may be used to sense spun yarn denier, or an optical device may be employed; however, as shown in Figure 9.10, most of these sorts of devices measure the capacitance of a moving fiber or yarn using electronic circuits. Output from the device may be used to control roll speeds or amount of draw or whatever is needed. Data collected by the device may be further analyzed

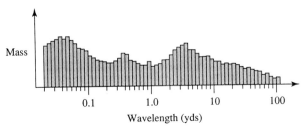

FIGURE 9.11 Spectrum analysis of output from Uster evenness tester on 10s cotton yarn. The results suggest that the yarn is characterized by mass fluctuations that occur periodically every 10 inches and a more severe one every 2.5 yds.

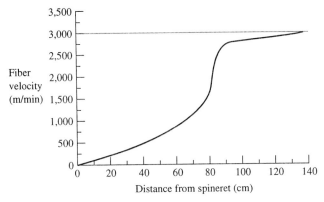

FIGURE 9.12 Velocity of poly(butylene terephthalate) fiber as a function of position in the extrusion line.

using a spectrum or frequency analyzer. Often the frequency of a perturbation can give information regarding the source of the perturbation, as shown in Figure 9.11. For example, if the wavelength of a pertubation in undrawn filament yarn is 10 m, then, largely from experience, we know that such variation must occur in the spinning (extrusion) line.

9.4.3 Fiber Velocity and Diameter Measurements

Another technique that can provide continuous information on fibers is the laser Doppler anemometer, which gives the velocity at any point of a moving fiber or yarn. The technique relies on the principle of frequency or Doppler shift associated with the velocity of particles, such as titania, trapped in a yarn. An example of time-averaged data from a laser Doppler anemometer is shown in Figure 9.12. These data were obtained during the extrusion and wind-up of poly(butylene terephthalate) at 4000 m/min. Under the conditions used, solidification occurs 80–90 cm from the spinneret. Several commercial velocimeters using the Doppler shift principle are available.

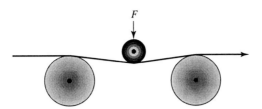

FIGURE 9.13   Fiber or yarn tensiometer.

9.4.4 Molecular Orientation Determination

To determine molecular orientation, which is important to modulus, strength, and elongation-to-break, a number of techniques may be used. Polarized light microscopy can be used to monitor retardance or birefringence with time, but sophisticated laser devices are more rapid and convenient to use. These sorts of instrumentation are generally in development stages only (see, for example, S. Kumar and R. S. Stein, *Journal of Applied Polymer Science* 34 (1987), 1693).

9.4.5 On-Line Measurement of Mechanical Properties

It is not possible to measure strength on-line, since strength testing is destructive; however, modulus may be measured in a number of ways. The load at a certain specified elongation may be monitored. The results of such a test are shown in Figure 9.9. Unfortunately, these devices cannot be operated at the speeds normally encountered in fiber processing. Hence, they are off-line techniques. Alternatively, pulse propagation techniques can be used to determine the sonic modulus.

9.4.6 Tension Devices

A number of devices for measuring threadline tension are available. All of the devices are contact devices, but sometimes the contact is minimal. Many instruments divert the fiber from its normal straight path and require the fiber to pass over a series of frictionless rolls, as shown in Figure 9.13. The tension is measured by the deflection on the central tension roll. At high fiber or yarn speeds the inertial corrections required can be significant.

9.5 Summary

Elastic modulus is an intrinsic property of a material. Measurements of fiber modulus give information on the structure of the material. On the other hand, strength and elongation are extrinsic properties. They give information on only a small volume of the material, the weakest portion.

All materials have statistical variations in strength. Variations exist from fiber to fiber and along a fiber. Peirce developed a model, the weakest-link model, that shows one way to determine the effect of variability on fiber and yarn strength. Weibull also developed a model to deal with the statistical nature of strength. Both theories are consistent with the observation that as test volume decreases, measured strength of a brittle material increases, as does the coefficient of variation.

The effect of twist on strength has been discussed. Yarn strength decreases with $\cos^2 \theta$, where θ is the helix angle; however, twist is necessary in staple yarns to ensure translation of force using friction from one fiber to the next. Stress transfer improves with twist. When stress transfer is not required, such as in continuous filament yarns, twist may not be beneficial.

The modulus of a fiber is derived from the strength of the bonds present in the macromolecules—the polymer chemistry—and the physical structure of the fiber—the molecular orientation. Axial modulus increases with molecular orientation, but the increase is at the expense of transverse modulus.

Several techniques that may be used to monitor or control variability in fibers, yarns, or perhaps fabrics are discussed. These range from the exotic—laser velocimetry, fiber capacitance, load at specified elongation, and fiber retardance—to the more mundane—pressure and temperature fluctuations.

References

W. D. Kingery, H. K. Bowen, and D. R. Uhlmann. *Introduction to Ceramics*, 2nd ed. New York: Wiley-Interscience, 1976.

W. E. Morton and J. W. S. Hearle. *Physical Properties of Textile Fibers*, 3rd ed. Manchester, England: The Textile Institute, 1993.

J. P. Schaffer, A. Saxena, S. D. Antolovich, S. B. Warner, and T. H. Sanders, *Materials Engineering*. New York: Times Mirror Books, 1995.

W. Watt and B. V. Perov, eds. *Strong Fibers*, V1, *Handbook of Composites*. New York: Elsevier, 1985.

Problems

(1) Explain why fiber staple length and yarn twist are important to yarn strength.

(2) Sketch a reasonable plot of cotton fiber denier as a function of position along a length of fiber for five different fibers, all from the same field.

(3) Sketch a reasonable plot for the denier of a fiber spun at 3000 m/min in which the extrusion pump showed a cyclic variation in speed every five seconds.

(4) The tenacity of a brittle fiber is measured at 1 cm test length and again at 10 cm. How will the strength and coefficient of variation of strength vary between the two test lengths?

(5) Why are very high modulus or tenacity fibers used in reinforcement applications *not* used as textile fibers?

(6) An experimentally observed fact is that as molecular orientation increases, the effect of defects on strength decreases. Recalling that oriented fibers are fibrillar, rationalize the observation.

(7) Two sets of fibers have the same tenacity at 1 cm test length, 10 g/d, but set A has a standard deviation in tenacity of 3 g/d and set B only 1 g/d. Calculate the tenacity of each set of fibers at 20 cm test length. A yarn is made from each set of fibers. Which yarn will be stronger at 20 cm with low twist? high twist?

(8) Suppose you work in a continuous filament polyester plant. There seems to be a chronic variation in fiber diameter. You hypothesize that the problem is a day/night problem, that is, that smaller fibers are made by day and larger ones by night. How would you proceed to prove your hypothesis?

(9) Soil samples may be taken by drilling with a hollow drill and examining the soil that is contained in the drill after the operation. The technique is called coring. Discuss whether coring or modified coring may be a viable technique for sampling fibers in large bales.

(10) Decide whether the following properties are intrinsic or extrinsic: electrical conductivity, dielectric strength, moisture regain, shear modulus, strain-to-fail, Poisson's ratio, and birefringence.

(11) A staple yarn is tested, and it is found that failure occurs by fiber slippage. No fibers break. An inflatable rubber device is patented that supposedly doubles the lateral force on the fibers without changing the fibers' alignment in the yarn. How will the yarn strength change with a doubling of the lateral force?

10

Effects of Time and Temperature on Mechanical Properties

We have learned about the mechanical response of fibers subjected to a variety of deformation schemes. Up to this point, however, we have treated deformation as though it occurs instantly in response to stress. While this is a good beginning, we must realize that it is only an approximation. We are ready to discuss the time-dependent behavior of textile and industrial fibers.

The fact that organic fibers show time-dependent mechanical behavior comes as no surprise. It is a characteristic of most polymers. In fact, it is a characteristic of *all materials* that are used or tested at temperatures that are sufficiently near the softening or melting temperature that some flow can occur on the time scale of the experiment. In general, materials show time-dependent deformation when their use or test temperature is 40 percent of the absolute melting temperature. Room temperature is sufficiently high that most polymers show viscoelastic—combined viscous and elastic—mechanical behavior under ordinary use conditions. The behavior may be tolerable, or it may not be, depending on the application. The time dependence of mechanical properties manifests in several ways.

10.1 Strain Rate and Temperature Dependence

When we test fibers with a fixed gauge length at various crosshead speeds or strain rates, we soon learn that the results depend on the conditions used for the test. Strain rate affects the stress–strain behavior of fibers.

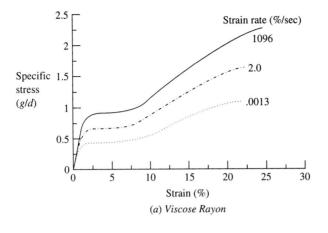

(a) *Viscose Rayon*

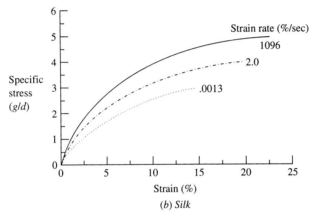

(b) *Silk*

FIGURE 10.1 Effect of strain rate on mechanical properties of viscose rayon and silk.

Strain rate is the rate of straining in a mechanical test. If the gauge length is x cm and the crosshead displacement is y cm/min, then the strain rate is

$$\dot{\epsilon} = d\epsilon/dt = \Delta l/l\Delta t = (\Delta l/\Delta t)/l = y/x$$

The effect of strain rate on the tensile properties of various fibers is shown in Figure 10.1. Hookean elastic materials show no strain rate sensitivity. Bond stretching and flexing are time-independent. Strain rate dependence is a consequence of the fact that polymers are not ideal *elastic* solids, but also contain a *viscous* component. Viscous materials are characterized by time-dependent deformation. Consider the effect of a shear stress applied to a fluid, as shown in Figure 10.2. When the fluid is confined between two plates and the upper plate is moved as shown, the motion produces not a shear strain, as was the case for a solid, but rather a velocity gradient normal to the flow direction. Shear stress is proportional to the velocity gradient. The proportionality

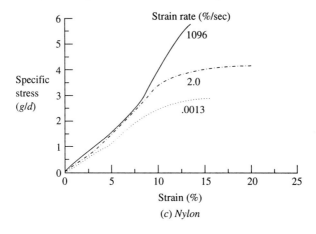

(c) Nylon

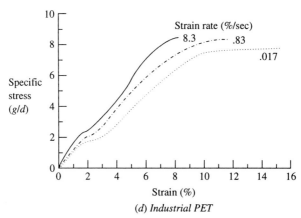

(d) Industrial PET

FIGURE 10.1 *(Continued)* Effect of strain rate on mechanical properties of nylon and industrial PET fibers.

constant is viscosity. Shear stress,

$$\tau = \eta \, dv_x/dy = \eta d(dx/dt)/dy = \eta d(dx/dy)/dt = \eta d(\gamma)/dt$$

so that

$$\tau = \eta \dot{\gamma}$$

where

v_x = velocity in the x direction,

γ = shear strain,

$\dot{\gamma}$ = shear strain rate, and

η = viscosity, or (fluidity)$^{-1}$.

This equation is known as Newton's law of viscosity. Viscosity is the proportionality constant between shear stress and shear strain rate. Viscosity is a measure of energy dissipation upon shearing a fluid. The higher the viscosity, the more energy is dissipated during a given shear flow, so the more the fluid heats up. The units of viscosity are poise, g/cm-sec.

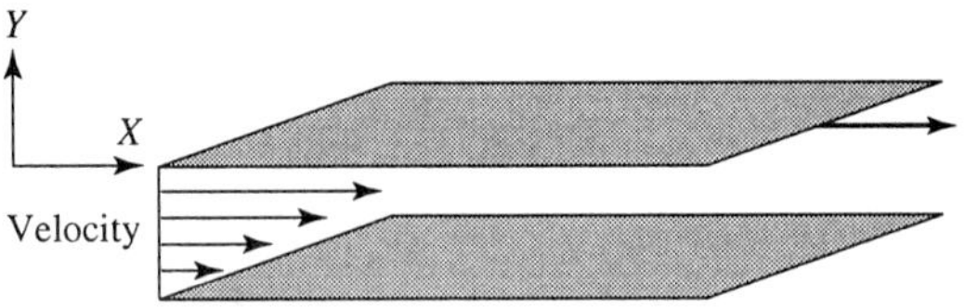

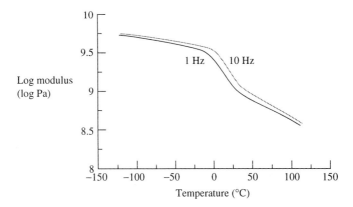

FIGURE 10.2 Experimental situation used to define shear viscosity.

FIGURE 10.3 Effect of temperature on modulus of isotactic polypropylene.

Another effect of the time dependence on mechanical properties can be seen by measuring the elastic, or Young's, modulus at various temperatures. A plot of modulus with temperature provides a wealth of information on polymers. The temperature dependence of the modulus of polypropylene, a semicrystalline polymer, is shown in Figure 10.3. The modulus decreases at the glass transition temperature, about 0 °C at 1 Hz. T_g is that temperature range where the polymer segments in the glassy regions have sufficient kT, or thermal energy, that motion becomes possible on the *time scale of the test*. Increasing the rate of the test causes the curve to shift right, as shown in Figure 10.3. Hence, there is an equivalence between temperature and strain rate, which is reciprocal time. The equivalence was quantified by Williams, Landel, and Ferry in what is now known as the WLF expression:

$$\log[t\{T\}/t\{T_g\}] = \frac{-17.44(T - T_g)}{51.6 + (T - T_g)}$$

where

T = temperature of test,

T_g = glass transition temperature or any reference temperature in the vicinity of the glass transition temperature, and

t = time.

Increasing strain rate produces the same effect as decreasing temperature or time. Substitution of $\log[t\{T\}/t\{T_g\}] = \log 10 = 1$ and solving for $T - T_g$ shows that the curve shifts right roughly 3 °C per decade increase in strain rate. The value of the WLF expression should not be overlooked. It states that time–temperature superposition is the same for all polymers, regardless of composition. (The WLF expression is limited to $T_g \leq T \leq T_g + 100$ °C.)

We have now seen two effects of the time dependence of mechanical properties. The key to understanding viscoelasticity, the time dependence of mechanical behavior, is to realize that polymers contain elements that behave in a viscous as well as elastic fashion. Thus, the σ–ϵ curve of a fiber varies according to the ability of the various components of the fiber to respond to the stress in the time scale of the experiment. As strain rate increases, the fluid elements behave more and more like solids. This behavior can be readily demonstrated using silly putty. When silly putty is strained at a low rate, it draws out like soft chewing gum or a material with a low modulus. When strained at a high rate, such as being bounced, it behaves like an elastic solid or a material with a high modulus. Increasing the temperature is equivalent to decreasing the strain rate. The effect is quantitative, and it is called time–temperature superposition.

10.2 Creep and Stress Relaxation

Creep and stress relaxation are the names of tests that have been developed to probe the time dependence of mechanical properties. Either test may be used, and equivalent information is learned from each test. The test techniques used to show and quantify the time-dependent properties of materials are described in Figures 10.4 and 10.5.

10.2.1 Creep

Attach a load to a sample and measure the length or strain with time. The stress produces an immediate elastic response, which is followed by a time-dependent viscous response. The elastic response is determined by the load applied, the cross-sectional area, and the modulus of the material. The viscous response is a measure of the time-dependent deformation. Polymers may creep to failure or, depending on polymer, stress, temperature, etc., may creep to a point and then effectively halt. Next we show the experimental setup for stress relaxation and the response of a fiber.

10.2.2 Stress Relaxation

Take a sample, length l, and extend it to a length $l + \Delta l$ and hold. Measure stress over a period of time, as shown in Figure 10.5. Typically the stress decays to a value somewhat larger than zero.

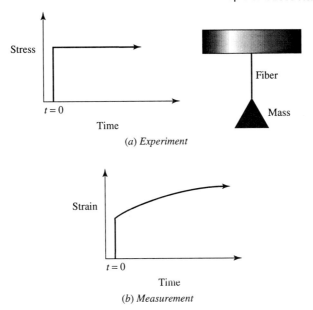

FIGURE 10.4   Creep testing and response of a viscoelastic fiber.

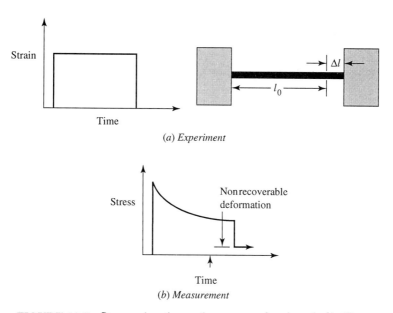

FIGURE 10.5   Stress relaxation and response of a viscoelastic fiber.

TABLE 10.1   Creep Behavior of Fibers

Structural Feature	Effect on Creep Behavior
Increased crystallinity	Decrease
Increased glass transition temperature	Decrease
Increased molecular orientation	Decrease
Increased molecular weight	Decrease
Increased crosslinking	Decrease
Increased molecular polarity	Decrease

Do other materials show creep and stress relaxation behavior? Yes, generally, when the use temperature is 40 percent times the absolute melting temperature, creep and stress relaxation are significant. For example, consider aluminum. T_m is approximately 660 °C = 660 + 273 = 933 K. 40% × 933 K = 373 K = 100 °C. Aluminum is one of the lowest-melting-temperature metals, yet it does not creep significantly at room temperature. Indeed, most metals and ceramics creep only at elevated temperatures. Consider now the example, PET. T_m is approximately 255 °C = 255 + 273 = 528 K. 40% × 528 K = 211 K = −62 °C. Most polymers creep at room temperature. Creep and stress relaxation can be serious problems for polymeric materials. Often, time-dependent deformation precludes a polymer's use in certain applications. Any application that involves a continuous load or strain is one that requires materials with good creep or stress-relaxation behavior. The factors that affect creep tendencies of various fibers are summarized in Table 10.1. Because of its chemical architecture and microstructure, Kevlar® shows excellent creep resistance. Polyethylene and polypropylene have notoriously poor creep resistance. An example of the creep behavior of polyethylene is shown in Figure 10.6.

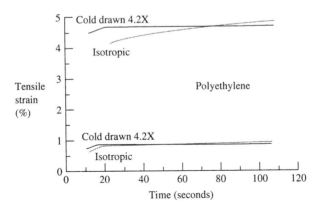

FIGURE 10.6   Creep behavior of polyethylene under low and high load. (M. W. Darlington and D. W. Sanders. *Journal of Physical D: Applied Physics* 3 (1970), 535.)

10.3 Models of Viscoelastic Behavior

The viscoelastic behavior of polymers can be modeled using mechanical elements. The mechanical models provide a useful qualitative picture of viscoelastic behavior, but they do *not* do a good quantitative job of modeling behavior. Elastic materials are modeled with springs, —⋀⋀⋀— , and viscous materials are modeled with dashpots, —▮▯— . Viscoelastic materials are modeled with both springs and dashpots, as shown in Figure 10.7. Springs show ideal elastic behavior (Figure 10.8):

$$\sigma = E\epsilon$$

Dashpots show ideal viscous behavior (Figure 10.9):

$$\tau = \eta\dot{\gamma}$$

(*Note: You will often see $\sigma = \eta\dot{\epsilon}$ for a dashpot. When discussing the behavior of a fluid, this expression is technically incorrect; however, when discussing the macroscopic behavior of the dashpot, tensile stress and strain are acceptable.*)

Let us now consider the behavior of a *Maxwell* element to loading in *stress relaxation*. The governing equation for its behavior may be developed by considering the total strain and the strain in each element:

$$\epsilon_{total} = \epsilon_{elastic} + \epsilon_{plastic}$$

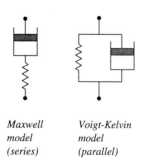

Maxwell
model
(series)

Voigt-Kelvin
model
(parallel)

FIGURE 10.7   Mechanical models for viscoelastic behavior of fibers.

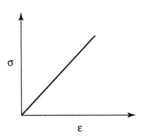

FIGURE 10.8   Statement of Hooke's law.

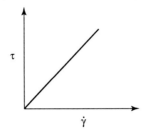

FIGURE 10.9   Statement of Newton's law of viscosity.

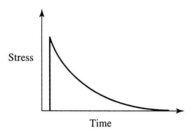

FIGURE 10.10   Behavior of a Maxwell element in stress relaxation.

Taking the time derivative of each term,

$$\dot{\epsilon}_{total} = \dot{\epsilon}_{elastic} + \dot{\epsilon}_{plastic}$$

so that after substitution for the behavior of the ideal elements,

$$\dot{\epsilon}_{total} = \dot{\sigma}/E + \sigma/\eta$$

This is the equation governing motion in the Maxwell model. In the particular case we are considering, stress relaxation, the length is invariant. Hence, the total strain and total strain rate are zero in stress relaxation:

$$\dot{\sigma}/E + \sigma/\eta = 0$$

Solving for σ,

$$d\sigma/\sigma = (-E/\eta)\,dt$$

so

$$\ln\sigma = (-E/\eta)t$$

and, hence,

$$\sigma = \sigma_o \exp[(-E/\eta)t] = \sigma_o \exp[-t/\tau_R]$$

where

$$\tau_R \equiv \eta/E$$

the relaxation time. The behavior of a Maxwell element in stress relaxation is sketched in Figure 10.10. Compare the result with that of a real material, shown in Figure 10.5. Clearly the agreement is reasonable, in that stress decreases nonlinearly with time, but not quantitatively correct. Real materials cannot be described by a single relaxation time, τ_R. Mechanical models for viscoelastic behavior are imperfect:

- The Maxwell model has permanent set in creep. The Voigt–Kelvin model does not. Polymers have permanent set.
- The Maxwell model continues to creep without limit. The Voigt–Kelvin model has limits. Fibers may do either, but they usually creep without limit.

Qualitatively the models are useful. A good feel for the models can be achieved by examining the quick and slow response of the two models:

- Springs stretch only to a finite limit. Dashpots can extend without limit.
- Springs show no time dependence.
- The dashpot is "hard" when the strain rate is high.
- The dashpot is "soft" when the strain rate is low.

As an exercise, return to the behavior of silly putty and explain its behavior at both high and low strain rates, perhaps using the models. At low strain rate silly putty deforms in a fluidlike fashion, requiring low stresses and responding with large strains. At high strain rate the deformation more nearly parallels that of an elastic solid and failure is brittle. As a further exercise, develop the equation governing the time-dependent strain of the Voigt–Kelvin model in creep. Begin with an equation for the total stress (see problem 18).

10.4 Dynamic Mechanical Analysis

Dynamic mechanical testing is designed to assess the viscoelastic behavior of polymers. In this technique a small sinusoidal strain is applied to a material, $\epsilon = \epsilon_o \sin \omega t$. How does the material respond?

An elastic material responds immediately with a stress, according to Hooke's law:

$$\sigma = E\epsilon = E\epsilon_o \sin \omega t$$

The response, stress, is in-phase with the applied strain.

A viscous material responds with a strain rate, according to Newton's law:

$$\tau = \eta = \eta \dot{\epsilon}_0 \omega \cos \omega t$$

The response to the stress, strain rate, is 90° out-of-phase with the strain. The stress lags the strain.

A polymer fiber responds intermediate to these two limiting cases:

$$\sigma = \sigma_o \sin(\omega t + \delta)$$

The stress lags the strain by an angle δ. For an elastic solid such as a ceramic, $\delta = 0°$. For a fluid, $\delta = 90°$. In polymer fibers, $0° < \delta < 90°$, as shown in Figure 10.11.

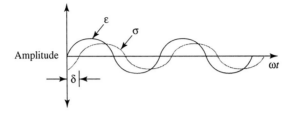

FIGURE 10.11 Sinusoidal strain and stress in fibers.

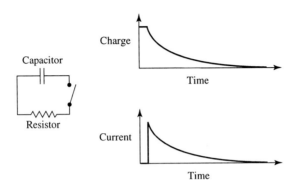

FIGURE 10.12 Typical RC circuit. Switch is closed and current flows until capacitor is discharged. Charge decays to $1/e$ of its original value in time RC.

For those of you who have completed a course in electricity and magnetism, an electrical analog may be helpful. Consider the case of an RC (resistance–capacitance) circuit. The equations governing the behavior of such a circuit are

$$\text{Resistor} \qquad I = V/R$$
$$\text{Capacitor} \qquad I = C(dV/dt)$$

The current, I, flowing through a resistor is in-phase with the voltage, V. The current flowing through a capacitor is 90° out-of-phase with the voltage. In an actual RC circuit the phase angle varies between these two limits, as shown in Figure 10.12. Note that the product RC has units of time. It is the time constant, fully equivalent to τ_R for viscoelastic behavior.

Let's return to the case of dynamic mechanical analysis and fiber behavior. Our intention is to address another point that has been heretofore ignored. As shown in Figure 10.13, modulus actually consists of a real part and an imaginary part:

$$E^* = E' + iE''$$

where

$E^* =$ complex modulus,
$E' =$ storage modulus, and
$E'' =$ loss modulus.

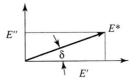

FIGURE 10.13 Modulus in complex space showing definition of δ.

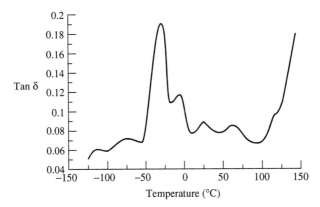

FIGURE 10.14 Tan δ in polypropylene with temperature. (G. M. Bartenev and R. M. Aliguliev. *Vysokomol. soyed* A26 (1984), 1236–1245.)

In many materials the loss modulus, E'' is small and is safely ignored. This is not true with many polymers or other materials as they begin to soften. The complex modulus can be plotted in complex space, showing that the tangent of the loss angle,

$$\tan \delta = E''/E'$$

Tan δ is a useful, meaningful quantity. It is a measure of the amount of energy lost to that stored or returned. When tan δ is high, the fiber is absorbing mechanical energy and converting it to heat. Mechanical energy may be absorbed by the various parts of the chain or chemical structure, analogous to the way a violin string can be set into resonance by an appropriate frequency sound, or the Verrazano-Narrows bridge set into oscillation by the appropriate winds. When tan δ is plotted as a function of temperature, a curve such as the one shown in Figure 10.14 for PP is developed. The peaks in tan δ represent temperatures at which the polymer absorbs a significant portion of the energy input. There are major peaks at the melting temperature, 165 °C, and the glass temperature, −30 °C. Other peaks occur at temperatures that depend on the details of the chemical structure of the chains.

Tan δ may be determined using a torsion pendulum. The apparatus was described in Chapter 8. There we showed that the shear modulus, G, is a function of the frequency, geometry, and the fiber properties:

$$G = 2lM\omega^2/\pi R^4$$

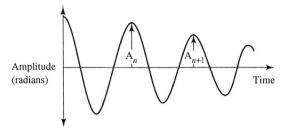

FIGURE 10.15 Data from torsion pendulum.

Tan δ is a measure of the viscous losses in a material or the energy absorbed by the material. A material with high loss will damp oscillations quickly. Hence, in a given setup, tan δ is determined solely by the damping rate of the pendulum. If the amplitude between successive oscillations is A_n and A_{n+1}, then the logarithmic decrement, Λ, as indicated in Figure 10.15, is

$$\Lambda = \log(A_n / A_{n+1})$$

It can be shown that for small strains,

$$\tan \delta = \Lambda / \pi$$

A torsion pendulum placed in an oven or cryogenic environment can be used to obtain data over a range of temperatures. Still better is to use a commercially available dynamic mechanical analyzer. In general, these instruments measure the stress lag directly when a sinusoidally varying strain (superimposed on top of a small static strain) is applied to a sample. The instruments are designed to be operated from roughly -100 °C to 350 °C. Dynamic viscometers are available for work on melts, providing information on $\eta^* = \eta' + i\eta''$. In addition, analogous dynamic electrical instruments are available, providing information on the relative dielectric constant, $\kappa^* = \kappa' + i\kappa''$.

Let us return to the behavior observed in the first figure of this chapter. Why do σ–ϵ curves show higher modulus, higher strength, and lower strain-to-fail with increasing strain rate? As strain rate increases, the molecules have less time to respond. Molecules that may have acted in a viscous fashion at lower strain rates or higher temperatures now act like elastic elements. Consequently, the modulus increases. Since there is less creep occurring during the test, the strength increases and the strain-to-fail decreases.

10.5 Sonic Modulus

Young's, or the elastic, modulus can be determined using a mechanical testing machine or a dynamic testing machine that, as described previously, imposes a dynamic strain and measures a dynamic stress. In addition, modulus may be assessed using sonic vibrations, as shown in Figure 10.16. Two transducers

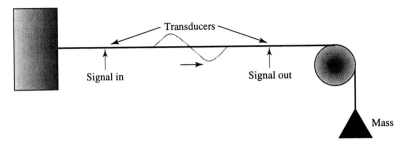

FIGURE 10.16 Schematic of pulse propagation instrument for measuring modulus in fibers.

are used. A sonic pulse is put into the fiber with one transducer. The other transducer is a specified distance from the first, and it detects the sonic pulse. The sound wave velocity, c, is determined using simple physics:

$$c = \text{distance/time}$$

It can be shown that the square of the sound wave velocity is proportional to the specific modulus of the material:

$$c^2 = E/\rho$$

The sonic modulus is a measure of the high-frequency modulus. Hence, it is equivalent to that measured at very high strain rates. It is generally greater than a modulus value measured using a mechanical testing machine.

10.6 Summary

Up to this chapter in the text we had led you to assume that stress–strain curves are independent of the rate at which the data are procured. This is true if and only if the material is in a temperature and strain rate range in which properties are time-independent. Most polymers show time-dependent, or viscoelastic, behavior at room temperature. The time effects can be significant and can bear on potential applications of fibers.

We probe the time-dependent behavior using creep or stress relaxation tests. The tests are different, but they provide equivalent information. In a creep test the strain increases with time in a sample under constant load. In the stress relaxation test the stress decays with time after the sample is given an instantaneous strain.

Models are too often used to develop mathematical relations that quantitatively define creep and stress relaxation data. The models consist of elastic elements, springs, and viscous elements, dashpots. Combined in series, the two-element model is the Maxwell model. In parallel, the two-element model is the Voigt–Kelvin model. Neither model accurately portrays the behavior of a polymer, but they provide a valuable means to visualize and conceptualize the behavioral response of viscoelastic materials.

The complex modulus shows both the time-independent and time-dependent parts of the modulus. Tan δ is a measure of the relative amount of time-dependent deformation. When δ is 0°, the fiber shows ideal elastic behavior. When δ is 90°, the fiber is an ideal fluid. Tan δ can be measured using a torsion pendulum or with specially designed instruments called dynamic mechanical analyzers.

The elastic, initial, or Hookean modulus is usually measured using a mechanical testing machine; however, pulse propagation techniques may be used. The frequency of the sound waves demands that the sample respond quickly. Hence, the sonic modulus is usually larger than that measured using a mechanical testing machine at high strain rates.

References

J. D. Ferry. *Viscoelastic Properties of Polymers*, 3rd ed. New York: Wiley, 1980.

G. Grover, S. Zhu, and I. C. Twilley, *Textile Research Journal* 63 (1993), 257.

W. E. Morton and J. W. S. Hearle. *Physical Properties of Textile Fibers*, 3rd ed. Manchester, England: The Textile Institute, 1993.

F. Rodriguez. *Principles of Polymer Systems*, 3rd ed. New York: Hemisphere, 1989.

I. M. Ward. *Mechanical Properties of Solid Polymers*, 2nd ed. New York: Wiley-Interscience, 1983.

I. M. Ward, ed. *Structure and Properties of Oriented Polymers*. Essex, England: Applied Science Publishers, 1975.

Problems

(1) A filament of nylon is stretched to 80% of its limit and held at constant length for 10 days. Assuming the filament does not fail, sketch how you anticipate the stress will behave over the 10 days, including the initial stretch and the final release.

(2) Use Figure 10.1 to calculate the ratio of modulus of nylon that is strained at 1096%/sec to that strained at 0.0013%/sec.

(3) Why does relative humidity change the tenacity, modulus, and elongation of cotton as well as the time dependence of these properties?

(4) A block copolymer is synthesized. A single chain consists of a block of polystyrene, a block of polybutadiene, and a block of polystyrene. The polystyrene block has a glass transition temperature of 100 °C, and the polybutadiene has a glass transition temperature of $\ll$ 0 °C. The sample has no crystallinity. The polymer consists of domains of polystyrene in a sea of polybutadiene. Describe the mechanical behavior of the polymer at room temperature.

(5) Using curves for cotton and polyester, describe how temperature and strain rate affect the stress–strain curve of textile fibers.

(6) Polymers are often modeled as springs and dashpots. Why? What physical region in the microstructure of the polymer corresponds to the spring and what region corresponds to the dashpot?

(7) Does the Maxwell model completely recover from strain? the Voigt–Kelvin model? Which is more realistic?

(8) List two applications for which a polymer fiber needs
   (a) excellent creep resistance.
   (b) high elastic recovery.
   (c) low modulus.
   (d) high work-to-break.

(9) In dynamic mechanical testing we discuss how much the stress lags the strain. Discuss and show the extent of lag in
   (a) a perfectly elastic material.
   (b) a perfectly viscous material.
   (c) a real polymer.

(10) Give three examples of behavior that demonstrate that fibers are viscoelastic.

(11) Sketch the stress–strain curve for each of the models shown:

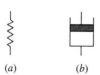

(a)   (b)

(12) Show that the Voigt–Kelvin model does not creep without limit.

(13) Water is forced through a small capillary immersed in a water bath. The experiment is repeated using honey, but the pressure is increased so the process is completed in the same time. Discuss differences in the final temperature of the water bath.

(14) While dynamic thermal analysis is generally used to probe the structure and properties of materials, $\tan \delta$ has been inversely correlated with the resiliency of carpet pile. Is this result reasonable? Do you anticipate that this fiber property is the only consideration in carpets? (Grover et al., 1993.)

(15) By about how much should the temperature be changed to nullify the effect of increasing the strain rate by a factor of 10^3?

(16) A fiber is known to be rather well represented by a Maxwell element. When a tensile stress of 3×10^3 Pa is applied for 15 seconds, the fiber strains 20%. When the stress is removed, the strain reduces to 10%. Calculate the relaxation time of the fiber.

(17) Calculate the axial velocity of sound propagation in a Spectra® fiber and compare it with that in an ordinary polyethylene fiber. The modulus of Spectra 1000® is 170 GPa and that of ordinary high-density polyethylene is about 1 GPa. The densities are 0.97 and 0.96 g/cm^3 respectively.

(18) Develop the equation that describes the time-dependent strain of the Voigt–Kelvin model subjected to creep. Begin with an expression for the total stress on the model.

(19) A polymer can be modeled reasonably well using a single Maxwell model with a fluid with viscosity 10^6 poise and a spring with modulus 10^6 Pa. If the system is loaded in stress relaxation, how long will it take for the stress to decay to $1/e$ of its original value?

11

Modeling Mechanical Properties

Now that we understand one class of physical properties, the mechanical properties, and the microstructure of fibers, we are in a position to begin to develop more accurate and quantitative structural models. A model for the structure of any fiber must be consistent with all the chemical and physical information available. The models may be used to calculate fiber mechanical properties on the basis of the properties and arrangement of the components. We may consider one component of a fiber to be the fibrils and the other the adhesive or glue that binds the fibrils. On the other hand, we usually consider the components of a fiber to be the noncrystalline and crystalline phases. Hence, we assume the structure of fibers is two-phase. This is a reasonable assumption for many fibers and a good first approximation for others. The three most useful two-phase models for fiber structure can be found in composite theory: the isostrain model, the isostress model, and the micellar model. We will investigate each one in turn. Finally, we introduce the next refinement in modeling, the three-phase model.

11.1 The Isostrain Model

Two continuous phases are aligned parallel with each other and stressed along the axial direction, as shown in Figure 11.1. In fiber theory we usually find it convenient to treat the crystalline phase as one component and the noncrystalline phase as the other component. In fiber-reinforced composite structures, the fibers are one component and the matrix is the other component. When a

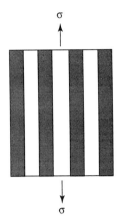

FIGURE 11.1 Isostrain model.

stress is applied as shown, the strain in all the elements must be equal:

$$\epsilon_c = \epsilon_1 = \epsilon_2$$

The load is divided according to relative modulus of the two components. The total force carried by the composite is

$$F_c = F_1 + F_2$$

where the subscripts refer to the composite, c, and the components, 1 and 2. Since $F = \sigma \times A$,

$$\sigma_c A_c = \sigma_1 A_1 + \sigma_2 A_2$$

where

A_1 = cross-sectional area of component 1, and

A_2 = cross-sectional area of component 2.

We are interested in the initial modulus, which is in the Hookean elastic region. Hence, $\sigma = E\epsilon$. Substitution gives

$$E_c \epsilon_c A_c = E_1 \epsilon_1 A_1 + E_2 \epsilon_2 A_2$$

Dividing by the strain and recalling that $\epsilon_c = \epsilon_1 = \epsilon_2$,

$$E_c A_c = E_1 A_1 + E_2 A_2$$

Dividing by the cross-sectional area of the composite,

$$E_c = E_1(A_1/A_c) + E_2(A_2/A_c)$$

Recognizing that the volume of component 1 is $A_1 \times l$ and the total volume is $A_c \times l$, the fraction A_i/A_c is the volume fraction of component i,

$$E_c = E_1 V_1 + E_2 V_2$$

Hence, the modulus of a two-phase material measured along the axis of the extended phase is a volume-weighted average of the two phases. Typically one component has a much higher modulus than the other component. If the volume fractions of the two components are comparable, one term on the right is small. Thus, the modulus of the composite is approximately the product of the volume fraction of the stiff component or phase and its modulus. For example, the modulus of the matrix is several orders of magnitude smaller than that of the fibers in fiber-reinforced polymer matrix composites. Hence, the term related to the matrix can be ignored, and the modulus of a fiber-reinforced polymer matrix composite is independent of the mechanical modulus of the matrix, a key conclusion.

11.2 Isostress Model

Envision the same structure as in the isostrain model, but this time apply stress *normal* to the fiber direction, as shown in Figure 11.2. Each element must carry the same load and stress, but the strain is partitioned according to the relative moduli:

$$\sigma_c = \sigma_1 = \sigma_2$$

Following the same sort of mathematics as in the isostrain case, it can be shown that the modulus in the isostress case is

$$1/E_{c\perp} = V_1/E_{1\perp} + V_2/E_{2\perp}$$

where the $\perp$ indicates that the modulus is measured normal to the axial direction. The modulus of polymer molecularly oriented along the fiber axis is much greater than that normal to the fiber length. Under these circumstances, the expression shows that the modulus of the composite is largely determined by that of the softer component or phase, as shown in Figure 11.3. Often the matrix is isotropic, so its properties do not vary with direction. On

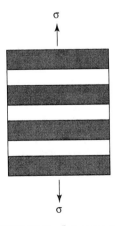

FIGURE 11.2   Isostress model.

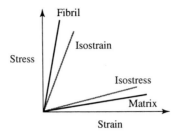

FIGURE 11.3 Schematic of the relationship of fiber modulus to those of the fiber components.

the other hand, fibers and fibrils may be highly anisotropic. Be sure to use the transverse modulus for isostress calculations. Recall that the transverse modulus is equal to the axial modulus in isotropic materials; however, in oriented polymers the transverse modulus may be as small as 1/100 of the axial modulus.

11.3 Micellar Model

The micellar model is a combination of the isostrain and isostress models just presented. In each of these models, both components are continuous, as shown in Figure 11.4. In the micellar model one phase is discrete. Assuming rectangular cells with dimensions shown, the modulus of the composite can be shown to be (Takayanagi et al., 1966)

$$E_c = \lambda[\phi/E_1 + (1-\phi)/E_2]^{-1} + (1-\lambda)E_2$$

where

$$\phi \times \lambda = \text{volume fraction of shaded phase}$$

This expression is useful for fibers with micellar structures, especially in fibers with discrete crystallites. In addition, it may be used on macroscopic solids

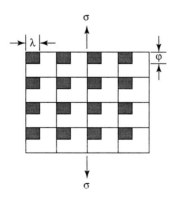

FIGURE 11.4 Micellar model.

to predict properties. An example showing how this model may be used in nonwoven fabrics is provided in problem 1.

11.4 Rule of Mixtures

In Chapter 3 we learned that the density, as well as a number of other properties, such as heat of fusion, heat capacity, and so on, do not depend on the spatial arrangement of the two phases, but rather only on the volume fractions. The value of density and these other properties, therefore, may be calculated using the rule of mixtures (ROM). Expressed in terms of density,

$$\rho_c = \rho_1 V_1 + \rho_2 V_2$$

Generalizing to two or more phases or components,

$$\rho_c = \sum_i \rho_i V_i$$

The arrangement of the two regions—isostress, isostrain, micellar, or other—does not affect this calculation.

Let us develop an important application of the two-phase model and introduce a few new concepts. Consider the net molecular orientation in a fiber and how it might be broken into component parts. One way we might proceed is to approximate a fiber as two-phase, crystalline and noncrystalline. We can use the rule of mixtures to establish an equation for fiber molecular orientation, and we will do that shortly. First let me say as a prelude to Chapter 12 that molecular orientation can be measured using X-ray diffraction and optical microscopy. X-ray diffraction can be used to quantify molecular alignment in crystals, and a term called *birefringence*, Δn, is a measure of the overall molecular orientation in a fiber. Now let us write an expression for fiber birefringence:

$$\Delta n = \Delta n_c V_c + \Delta n_a V_a$$

where the subscripts refer to properties in the crystalline and amorphous regions. The birefringence of a specific polymer fiber depends on its chemistry, molecular orientation, and chain packing. As a first approximation, birefringence is independent of the size, shape, and spatial distribution of the crystalline and noncrystalline phases. We use the symbol Δn^o to indicate the maximum value that a phase of polymer can achieve. It is called the intrinsic birefringence. An orientation factor is used to weight the intrinsic birefringence. For perfect orientation, $f = 1$. Hence,

$$\Delta n = (\Delta n_c^o f_c) V_c + (\Delta n_a^o f_a) V_a$$

This expression is valuable to fiber scientists. f_a is largely responsible for the ease of dye penetration, fiber shrinkage, and a host of other important characteristics, yet it is very difficult to directly measure f_a. At this time, I expect you to understand only that the preceding equation is simply an expression of the ROM for a two-phase material.

11.5 Refinements in the Models

The models just presented are the simplest models for fiber structure. They facilitate calculation of various physical and mechanical properties of fibers from a basic knowledge of fiber structure and component properties. An application of the models is provided in the chapter on optical properties, in which the molecular orientation is calculated on the basis of properties of the crystalline and noncrystalline phases. Another application of the ROM is the calculation of crystallinity based on heat of fusion or density presented in Chapter 3. Note that in our discussion we have at times mentally divided the fiber into fibrillar and interfibrillar material and used the isostrain model, as shown in Figure 11.5a. At other times we have mentally divided the fiber into crystalline and noncrystalline material and used either the isostress or lamellar (micellar) models, as shown in Figure 11.5b. These choices are not mutually exclusive, but offer complementary information.

The two-phase models presented are simplifications of real structures. They may be usefully employed in a number of situations for a variety of fibers. They are especially useful for providing qualitative information or order-of-

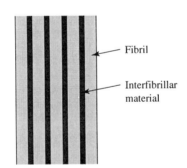

(a) Fibrillar model suggests use of isostrain analogy

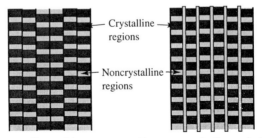

(b) Crystallite model suggests use of isostress or micellar analogies

FIGURE 11.5   Idealized models of fiber morphology.

FIGURE 11.6 Model of the structure of nylon fibers. (N. S. Murthy, A. C. Reimschuessel, and V. Kramer, *Journal of the Applied Polymer Science* 40 (1990), 249–262. Reprinted by permission of John Wiley & Sons, Inc. Copyright ©1990.)

magnitude estimates; however, the models cannot be used when accurate numbers are needed. We have, for example, ignored the fact that polymers often have thick interphase regions. We have ignored the presence of third or fourth phases. We also realize that oriented polymers may have regions of noncrystalline material that are highly oriented and other noncrystalline regions in the same sample that are characterized by low molecular orientation. In sum, two-phase models are approximations, and the calculations based on two-phase models should be treated as estimates.

At some point in your career, you will probably suffer the consequences of the breakdown of the two-phase model, in prediction of either properties or behavior of a fiber. At these times you may need to invoke more complex morphological models. The next most sophisticated model, and one that more accurately portrays the structure of industrial fibers, is the model that consists of three components: a crystalline phase, an interlamellar noncrystalline phase, and an interfibrillar noncrystalline phase. A model for such a fiber structure is shown in Figure 11.6. The interlamellar phase typically has lower molecular orientation than the interfibrillar phase. Consequently, solvent and dye can readily penetrate these regions. The difference in structure from fiber to fiber is modeled by rather small but significant differences in the details of the model, such as the amount of material in each region, the connectivity of the lamellae, etc. While we will not use this model in this text, it is a very good representation of fiber structure, and you will certainly learn more about it in an advanced fiber science course or in your reading.

11.6 Summary

In this chapter we have presented several models that are useful for calculating physical and mechanical properties of fibers that are based on the structural arrangement of the components present, or vice versa. We have limited our modeling to two-component structures, but more components may be easily incorporated. A two-phase morphology—crystalline and noncrystalline—is a reasonable first approximation of most fiber structures, as shown in Chapter 3. The major problems with two-phase models are that the properties of the noncrystalline regions are not constant, and other phases may be present, for example, thick boundaries between phases in the fiber or voids. On the other hand, fibers may be viewed as aggregates of fibrils given coherence by interfibrillar material. Again, a two-phase model can be invoked as a first approximation. When two-phase models break down, three-or-more-phase models may be employed, but then the simplicity inherent to a two-phase model is sacrificed. The models presented are useful not only for fibers, but also for bulk polymers, composite materials, thermally bonded nonwoven fabrics, etc. They are general.

References

K. K. Chawla. *Composite Materials*. New York: Springer-Verlag, 1987.

N. S. Murthy, A. C. Reimschuessel, and V. Kramer. *Journal of the Applied Polymer Science* 40 (1990), 249–262.

D. C. Prevorsek, P. J. Harget, R. K. Sharma, and A. C. Reimschuessel. *Journal of Macromolecular Science* B8 (1973), 127.

J. P. Schaffer, A. Saxena, S. D. Antolovich, S. B. Warner, and T. H. Sanders. *Materials Engineering*. New York: Times Mirror Books, 1995.

M. Takayanagi, K. Imada, and T. Kajiyama. *Journal of Polymer Science*, Part C, 15 (1966), 263–281.

Problems

(1) What information do you need to use the models of structure to facilitate calculation of the modulus of a nonwoven fabric with the bond pattern shown?

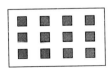

(2) Calculate the modulus of a uniaxial carbon fiber composite both normal and parallel to the fiber axis. The fiber content is 70 volume %. In the absence of better values, you may approximate the modulus of the epoxy to be 1/500 that of the fiber along the fiber axis and 1/10 that normal to the fiber axis.

(3) The density of a certain special polymer is approximately independent of the level of crystallinity. The tensile modulus of a fiber of the polymer is 300 g/d, and it is about 50% crystalline. The crystals are fibrillar (along the fiber axis), and the modulus of a fibril is on the order of 500 g/d. Estimate the axial modulus of the noncrystalline material. Can you estimate the transverse modulus of the noncrystalline material?

(4) A block copolymer is synthesized. A single chain consists of a block of polystyrene, a block of polybutadiene, and a block of polystyrene. The polystyrene block has a glass transition temperature of 100 °C, and the polybutadiene has a glass transition temperature of $\ll 0$ °C. The sample has no crystallinity. The polymer consists of (a) domains of polystyrene in a sea of polybutadiene or (b) domains of polybutadiene in a sea of polystyrene. How might the polymer behave at room temperature?

(5) The mechanical properties of a new fiber, which is difficult to grip in a testing machine, are required. The fibers come from nature and are present as one component in a two-component composite material. The fibers are aligned in one direction. From solvent studies it has been confirmed that about 60% of the composite is fiber. Mechanical testing of the composite in the axial direction shows that the modulus and strength are 300 and 22 g/d. The density of the composite is 1.40 g/cm^3. Mechanical testing of the isotropic matrix shows that the modulus and strength are 50 and 5 g/d. The density of the matrix is 1.35 g/cm^3. Estimate the fiber

(a) modulus.

(b) density.

(c) strength.

(6) Write an expression for the density of an imperfect composite containing fiber, matrix, and void. Write an expression for the density of a fiber that consists of three phases: crystalline, amorphous 1 (low orientation), and amorphous 2 (high orientation).

(7) Suppose a nonwoven fabric is made from a number of identical very long, straight fibers. In one case, all the fibers are parallel, essentially forming a yarn. Write an expression for the modulus of the yarn. In another case, the fibers are laid down flat but their orientation is random. Write an expression for the modulus of the fabric, assuming fibers loaded in compression will buckle and offer no resistance to loading.

PART THREE

Physical Properties

12

Optical Properties

The optical properties of fibers can be extremely important. Reflectivity, absorptivity, color, refractive index, and birefringence are some of the optical properties that may be used to characterize a material or may be of critical importance in applications of fibers. We will discuss optical phenomena as properties in service environments and as characterization tools for fibers. Any discussion of optical properties always reminds me of the stories I've heard about nylon women's bathing suits that were developed about 50 years ago. The new suits were gloriously advertised and anxiously awaited. They were to revolutionize the industry. The suits looked fine on the models and the photographs were appealing. Little did the manufacturers think about the fact that the fibers in the suits scattered light from the fiber's outer surface only. To the surprise of all, the wet suits were nearly transparent! This story alone demonstrates how important it can be to understand optical phenomena.

The key to understanding the intrinsic optical, dielectric, magnetic, and electrical properties of materials is to develop a mental picture of polarizability, the ability of electromagnetic radiation or any energy field to polarize material. Polarizability is the common thread uniting these properties, and they are all related as a consequence. Understanding polarizability will lead us to anticipate that dielectric materials, or materials with low electrical conductivity, are thermal insulators; that electrical conductors are opaque; and so on. In this chapter we focus on optical properties of fibers, but many of the concepts apply the other properties and other materials as well.

12.1 Polarizability and Refractive Index

Polarization is the state in which a material shows separation of charge, either permanently or momentarily. There are four mechanisms of polarization, shown in Figure 12.1. Electronic polarization is due to temporary distortion of an electron cloud. Ionic polarization results from a slight displacement of opposite charges. Molecular polarization occurs in polar molecules that can align in a field. Space-charge polarization requires redistribution of ionically charged layers. Light is electromagnetic energy or radiation that attempts to polarize transparent materials as it passes through them. Because visible light is relatively high-frequency radiation, only electronic mechanisms can respond sufficiently rapidly to affect the properties of the material and the light as it passes through the material and, hence, determine n, refractive index. The other polarization mechanisms involve motion of larger, heavier entities, and hence, they cannot respond to the rapid frequency cycling characteristic of visible light. Thus, the other mechanisms are inactive in visible light.

Refractive index has been defined in your physics courses, and we begin with their definition:

$$n = \sin i / \sin r$$

where

i is the angle of incidence, and

r is the angle of refraction, as shown in Figure 12.2.

Another definition of refractive index is

$$n = v_{\text{vacuum}} / v_{\text{medium}}$$

where v is velocity of light. Since light travels more slowly in a medium than in a vacuum, $n \geq 1$.

Refractive index varies with temperature and wavelength. The values given in most tables are at a specific temperature, usually 20 °C, and at a specific wavelength, often 589 nm. One problem associated with the optical microscope is called chromatic aberration. It is a result of the variation in refractive index with wavelength. The lenses in inexpensive optical systems are ground and positioned so that only one wavelength can be focused at a time. This causes the edges of a magnified image to show a colorful blur. More expensive objectives (apochromats) allow for the focusing of three wavelengths at once, giving a crisper image. The variation of refractive index of a material with wavelength is called dispersion. Dispersion causes white light to split into its color components as it passes through a prism.

One characteristic of refractive index that we know immediately is that it increases with density. The reason is rather simple: Electronic polarizability increases as the number of molecules or electrons per unit volume increases.

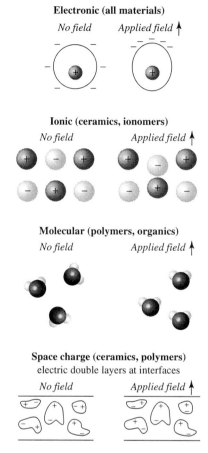

FIGURE 12.1 Polarization mechanisms in materials.

FIGURE 12.2 Refraction of an electromagnetic ray in a transparent medium.

The result is expressed as Gladstone and Dale's law:

$$(n - 1)/\rho = \text{constant}$$

The expression is useful when considering the influence of moisture on the refractive index of cotton, wool, or other fibers. Cotton has a lower density at 65 percent RH than at 0 percent RH (1.52 versus 1.55 g/cm^3), and wool has a higher density (1.31 versus 1.30 g/cm^3.)

12.2 Birefringence

In an isotropic material the refractive index is the same in all directions. When the material is anisotropic, such as a crystal or an oriented polymer, n varies with direction. Most fibers can be completely characterized by two macroscopic refractive indices, one parallel to the fiber axis, n_a, and one normal to the fiber axis, n_r, as indicated in Figure 12.3. (We assume the circumferential equals the radial refractive index in most fibers.) The subscripts a and r indicate the vibration direction of the radiation—axial or radial. The propagation direction is not important to birefringence. The difference in refractive indices, $n_a - n_r = \Delta n$, is called the birefringence, which is a measure of the difference in polarizability of the media to light vibrating parallel and perpendicular to the fiber axis. In general, the atoms along the backbone of a polymer are more or less polarizable than are the atoms in the lateral direction. Hence, birefringence may be used to assess molecular orientation in most fibers. The birefringence of polymer fibers is generally positive, a result of n_a exceeding n_r.

Fibers with different compositions have different polarizabilities. Hence, they cannot be readily compared. For example, the birefringence of a well-oriented PET fiber is much larger than that of a well-oriented PE fiber. *Do not* attempt to compare the molecular orientation of different fibers using raw birefringence data. Molecular orientation comparisons between fibers of different composition can be done using orientation functions, such as Herman's, which is discussed later in this chapter. Table 12.1 gives refractive indices and typical birefringences of various fibers. Even though semicrystalline fibers contain anisotropic crystallites, the crystallite orientation is essentially random in an unoriented fiber. Therefore, an unoriented fiber is characterized by a refractive index that is invariant with direction of measurement:

$$n_a = n_r = n_{\text{isotropic}}$$

Oriented fibers obey the conservation law:

$$n_{\text{isotropic}} = (n_a + 2n_r)/3$$

This equation may be useful when only one refractive index of a fiber can be readily measured. The birefringence may then be estimated using n_{iso}. An example is the fiber Kevlar®, which has such a high birefringence that n_a is outside the range of most standard refractive index fluids.

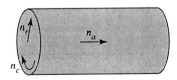

FIGURE 12.3 Refractive indices of fibers.

TABLE 12.1   Refractive Indices of Fibers

Fiber	n_a	n_r	Δn
Cotton[a]	1.578	1.532	0.046
Ramie and flax[a]	1.596	1.528	0.068
Viscose rayon[a]	1.539	1.519	0.020
Triacetate[b]	1.474	1.479	−0.005
Wool[b]	1.553	1.542	0.010
Silk[b]	1.591	1.538	0.053
Vicara (zein)[c]	1.536	1.536	0.000
Nylon[c]	1.582	1.519	0.063
PET	1.720	1.540	0.180
Acrylic (Orlon®)[c]	1.500	1.500	0.000
Acrylic (Acrilan®)[c]	1.520	1.524	−0.004
Polyethylene	1.550	1.510	0.040
Polypropylene	1.480	1.515	0.035
Kevlar® 49[d]	2.267	1.606	0.662
Glass (isotropic)	1.515	1.515	0.000

Sources: [a]J. M. Preston, *Transactions of Faraday Society* 29 (1933), 65; [b]J. M. Preston, *Modern Textile Microscopy*, London: Emmott, 1933; [c]A. N. J. Heyn, *Textile Research Journal* 22 (1952), 513; [d]S. Kumar, *Indian Journal of Fiber and Textile Research* 16 (1991), 52–64.

The relationship has also been used to determine the spiral angle of cotton, 27 to 34° (see Meredith, 1946), and other fibers. Alternatively, it can be used to determine the refractive index parallel and normal to the chains in fibers with molecules in a helical conformation, such as flax:

n(parallel to chain direction) = 1.595

n(normal to chain direction) = 1.531 (Meredith, 1946)

These refractive indices are of general use in that these are representative of the principal directions of cellulose in a highly crystalline form and, hence, may be used for all crystalline cellulosic samples.

12.2.1 Measurement of Refractive Indices and Birefringence

There are many ways to measure the two principal refractive indices or to measure birefringence directly. Perhaps the most common technique for mineralogists, which is also used by some fiber science microscopists, is to align a polarizer parallel to the direction of interest, so the electric field is vibrating, say, parallel to the fiber axis, and determine n_a. Then rotate the sample (fiber) 90° so the electric field is vibrating normal to the fiber and determine n_r. Refractive index is determined in each of the orientations by the Becke line technique:

   1. The fiber is immersed in a refractive index liquid and oriented so as to facilitate determination of one refractive index.

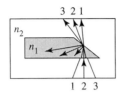
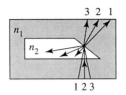

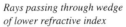

Rays passing through wedge   Rays passing through wedge
of lower refractive index   of higher refractive index

FIGURE 12.4   Becke line analysis.

2. The sample is focused and the stage then lowered and raised to determine whether the bright line near the edge of the fiber is inside the fiber or outside. The effect is shown in Figure 12.4 for the simpler case of a wedge-shaped object. If the bright line moves inside the fiber as the stage is lowered, then the fiber has a higher refractive index than the medium.

3. The fiber is removed from the liquid, cleaned (or a new section selected), and placed in another liquid. The process is repeated until the bright line is no longer observed. A perfect match means there is no refractive index change at the interface. The interface disappears and so too may the fiber.

The principle of refractive index matching can be readily demonstrated by placing a number of glass rods into a jar. Light is reflected and refracted by the glass rods. When the jar is filled with a refractive index liquid equal to that of the glass (1.515), such as vegetable oil, then the rods become invisible. This principle has been used industrially many times, such as in the formation of transparent composite sheets containing a polymer matrix and reinforcement fibers. These composites are used in solar glazing applications.

A compensator is the device most often used to determine birefringence of fibers. The principle of the compensator is nontrivial. Light incident on a fiber located in the 45° position between crossed polars, as shown in Figure 12.5a, divides the incident light into an ordinary ray and an extraordinary ray, as shown in Figure 12.5b. The ordinary ray has its electric field vibration parallel to the fiber axis. The extraordinary ray has its electric field normal to the fiber axis. The speed of the light varies according to refractive index, n_a or n_r. When the fiber is positively birefringent, the ordinary ray travels faster than the extraordinary ray. The difference in velocity gives rise to a difference in phase between the two rays. A compensator is inserted into the path of the ordinary ray, slowing the ray down, and adjusted to the point where the two rays are in-phase, as shown in Figure 12.5b. The amount of phase subtracted, or retardation, is proportional to the amount of phase difference set up by the difference in refractive indices. Birefringence may be calculated by

$$\Delta n = \text{retardance/thickness} = \Gamma/d$$

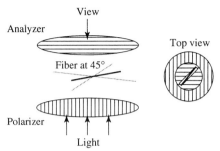

(a) *Light rays traveling through a fiber*

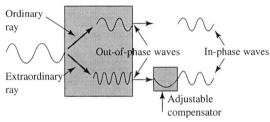

(b) *Measurement of fiber retardation with a compensator*

FIGURE 12.5 Measurement of birefringence in a fiber using the compensator technique.

for a round fiber with diameter d and retardance Γ. Fiber diameter is measured in a separate step essentially by projecting a calibrated grid onto the fiber silhouette.

A compensator cannot be easily used for nonround fibers. The Becke line technique or an interference microscope may be used. The interference microscope facilitates very accurate refractive index matching regardless of fiber cross-sectional shape. In the interference microscope, retardation bands are established normal to the fiber axis. When the bands do not deviate from their path as they travel through the fiber, the refractive index of the fluid and fiber are equal (H. R. E. Frankfort and B. H. Knox, US Patent 4,195,051, 1980). Many fiber scientists prefer to use interference microscopy to assess birefringence; however, few laboratories are equipped with interference microscopes, which can be rather costly.

Up to this point we have assumed $n_c = n_a$ and that refractive index does not vary with radial position. For most fibers these are indeed good approximations. For some wet-spun fibers, fibers melt-spun at high speed, or liquid crystalline fibers, the assumptions may break down. The interference microscope, again, may be used to determine whether refractive index varies along the radius of a fiber. We will not discuss the procedure in this text; however, there are numerous references in the literature, in part because graded index oxide fibers are a standard product in the optical waveguide industry. Let it suffice to say that when the fiber is immersed in a fluid selected so that part of the fiber has a higher index than the fluid and part of the fiber has a lower refractive index than the fluid, the retardation bands that traverse the fiber deviate both to the right and to the left as they pass through the fiber (H. R. E. Frankfort and B. H. Knox, US Patent 4,195,051, 1980).

12.2.2 Modeling with Birefringence

Modeling fiber structure was addressed in some detail in Chapter 11. We showed that a two-phase model may usually be used to gain some semiquantitative insight into fiber characteristics. The rule of mixtures was used to calculate a fiber property that is independent of the shape or positioning of the two phases, such as fiber density:

$$d = d_c V_c + d_a V_a$$

where

d = property,

V = volume fraction, and

a and c refer to crystalline and amorphous regions.

Birefringence is a measure of molecular orientation, regardless of whether the polymer is crystalline or amorphous. Birefringence is generally regarded as being independent of the size and shape of the two phases. (The latter approximation stems from assuming that form birefringence, which is very difficult to measure, is negligible. Even though we cannot verify this assumption, we will proceed.) Consequently, we can express birefringence as the sum of two terms:

$$\Delta n = V_c \Delta n_c + V_a \Delta n_a$$

In a two-phase structure $V_a = 1 - V_c$. Hence,

$$\Delta n = V_c \Delta n_c + (1 - V_c) \Delta n_a$$

The expressions for the birefringence of crystalline and amorphous regions can be broken down one step further:

$$\Delta n_c = \Delta n_c^o f_c \quad \text{and} \quad \Delta n_a = \Delta n_a^o f_a$$

where

$f_c = 1 - (3/2)\overline{\sin^2 \theta}$, an optical orientation function for the crystalline regions. θ is the angle of the molecule to the fiber axis. f_c is unity for a perfectly oriented crystal.

$f_a = 1 - (3/2)\overline{\sin^2 \theta}$, an optical orientation function for the noncrystalline regions.

Δn_c^o = maximum, or intrinsic, birefringence of crystalline polymer. Molecules in crystalline polymer are perfectly oriented and densely packed.

Δn_a^o = maximum, or intrinsic, birefringence of the noncrystalline regions.

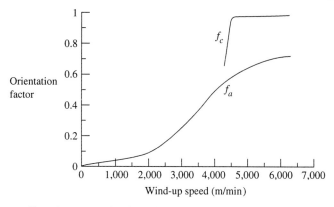

FIGURE 12.6 Development of orientation during spinning of poly(ethylene tereph-thalate.) (H. M. Heuval and R. Huisman, *Journal of the Applied Polymer Science* 22 (1978), 2229.)

Combining the equations allows us to assess the elusive f_a, the molecular orientation in the noncrystalline regions:

$$\Delta n = V_c \Delta n_c^o f_c + (1 - V_c) \Delta n_a^o f_a$$

f_a is a very important property of fibers. High f_a causes shrinkage at high temperature and leads to poor dye penetration. f_a is difficult to measure directly, but f_c can be measured using X-ray diffraction. $V_c = \chi$, crystallinity, and it can be measured using the techniques described in Chapter 3. Intrinsic birefringence can be calculated or found in the literature. When determined using the preceding equation, f_a is not accurate because of the difficulty of obtaining good values for intrinsic birefringence and the assumptions associated with the two-phase model; however, the estimates are nonetheless useful. Consider, for example, PET fiber spun at very high speed. At a take-up speed of about 4000 m/min, depending on molecular weight, fiber denier, and other variables, a decrease in the rate of increase in f_a is observed, as shown in Figure 12.6. This decrease is brought about by the onset of crystallization in the most oriented regions (A. Hamidi, A. S. Abhiraman, and P. Asher, *Journal of Applied Polymer Science* 28 (1983), 567). Hence, fibers spun above about 4000 m/min have unusual shrinkage and dye properties.

12.3 Reflection and Luster

The amount of light reflected from the surface and interior of a fiber is an important factor in many situations, as illustrated in Figure 12.7. In transmitted light microscopy, when a scientist needs to observe the interior structure of a fiber, surface reflection needs to be minimized. In reflected light microscopy, where a scientist needs to learn about the surface, reflection needs to be maximized. In most textile apparel applications, fibers need to cover the sur-

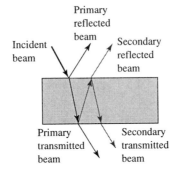

FIGURE 12.7 Interaction of light with a transparent medium.

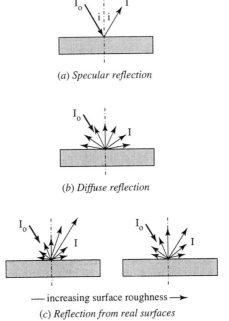

(a) Specular reflection

(b) Diffuse reflection

— increasing surface roughness ——➤
(c) Reflection from real surfaces

FIGURE 12.8 Types of surface reflection.

face as completely as possible and reflect light without glare. Light incident onto a surface will be reflected according to Fresnel's law. At normal incidence,

$$R = [(n - 1)/(n + 1)]^2$$

where

n = refractive index,

1 = refractive index of air, and

R = reflectivity, or the fraction of light reflected.

Trilobal   *Bilobal or (dogbone)*   *Octalobal*

FIGURE 12.9   Fiber cross sections.

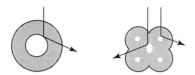

FIGURE 12.10   Light rays passing through hollow fibers.

Light may be reflected specularly, as shown in Figure 12.8a. In this case, all the light *leaving* the surface of the sample is well collimated at angle $r = i$. On the other hand, light may be reflected diffusely, as shown in Figure 12.8b. Here all the light leaving the surface goes off with equal intensity in all directions. Most real surfaces show a combination of specular and diffuse reflection, with a peak at angle r. Irregular or rough surfaces maximize diffuse reflection. Fiber shape is also an important factor. Hence, some synthetic fibers have unusual cross-sections, as shown in Figure 12.9. Fine fibers in general "cover" better than coarse fibers. (Cover is the ability of a fabric to hide the material beneath.) They have a higher surface area-to-volume ratio and hence, a higher surface area-to-weight ratio. In addition to scattering from the fiber surface, a portion of the light that enters a fiber is scattered. Scattering causes fibers to appear white, just as foamed polystyrene or foamed water appears white. The whitening is a desirable characteristic in many textile applications. To increase internal scattering, fine particles with a high refractive index may be added. Usually, fibers are in part delustered, meaning that specular reflection is made less obvious, with a few tenths of a percent of titania (TiO_2), which has a very high refractive index (2.71.) *Dull* or *semidull* is the jargon for delustered fibers. In addition, fibers may be hollow. The lumen has a low refractive index, that of air, causing refraction and scattering of the light as it enters and leaves the lumen, as illustrated in Figure 12.10.

Before moving on to discuss absorption, let me remind you of a fundamental conservation law of physics. All light incident onto the surface of a material is either reflected, transmitted, or absorbed. Hence,

$$A + R + T = 1$$

where

   A = fraction of light absorbed,
   R = fraction of light reflected, and
   T = fraction of light transmitted.

The amount of light scattered within a medium does not enter into this accounting, since all scattered light must later be either absorbed or transmitted.

12.4 Absorption and Dichroism

Metals are opaque because of the interaction of electromagnetic radiation in the visible region with electrons in the conduction band. Polymers, organics, and ionic solids are transparent if there is neither scattering nor heavy absorption. Density fluctuations (such as those facilitated by the presence of crystalline and amorphous regions or material and void regions), or perhaps more accurately, refractive index fluctuations, cause light scattering and, hence, opacity. An interesting example is Ivory® soap. Ivory® floats and appears white simply because of the many small air bubbles trapped in the soap. It is useful to remember that virtually all transparent polymers are noncrystalline, although not all noncrystalline polymers are transparent. Examples include polystyrene, polymethylmethacrylate, polycarbonate, and various thermoset or crosslinked resins such as epoxies and styrene-polyesters. A transparent ceramic may be noncrystalline or be one with very large crystals, perhaps single-crystal. Examples include the various oxide glasses and quartz, diamond, and other gemstones, which are all single crystals. In addition, if the crystals are sufficienty small, less than about 1/10 the wavelength of the incident light, then ceramics may be translucent or transparent. An example is Lucalox®, aluminum oxide with very small grains, which is the material that revolutionized high-intensity industrial outdoor lighting.

The color of an opaque object is defined chiefly by the wavelength of the reflected radiation. A green object is green because it absorbs all wavelengths in the visible spectrum except green. Green light is reflected to the eye. Similarly, a transparent or translucent green material viewed in transmission is green because it absorbs less radiation in the green region than in other visible regions. Green light reaches the eye. In general, being semicrystalline, textile fibers are translucent materials. Color is determined largely by the wavelength of the scattered rays. Absorption of light in materials can be made to occur at specific wavelengths, providing the material with well-defined and controllable color. For example, a green fiber may have the transmission curve shown in Figure 12.11. This result is achieved by adding chemicals or dyes that selectively absorb radiation. Lambert's law describes the amount of radiation absorbed:

$$I = I_o \exp(-\beta x)$$

where

I_o = intensity of incident beam,

β = absorption coefficient,

x = thickness of the penetration, and

I = intensity of light that has passed through a thickness of the material, x.

The absorption coefficient is proportional to the density of the dye added to the fiber or the dye concentration. Rewriting Lambert's law in terms of concentration gives the Beer–Lambert equation:

$$I = I_o \exp(-c_f x)$$

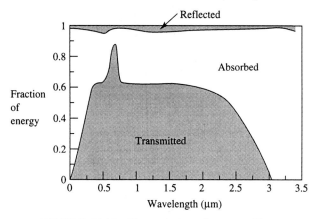

FIGURE 12.11 Transmission of a green fiber.

where

c = concentration, and

ϵ = extinction coefficient or absorption per unit of concentration.

Either of these expressions may be used to assess color characteristics and to characterize the absorption of light.

Light absorption characteristics may also be used to characterize the molecular structure of polymers. Rewriting Lambert's law:

$$\log(I/I_o) = -\beta x$$

The use of polarized light facilitates measurement of differences in absorption in oriented materials both along the fiber axis and normal to the fiber axis. When polarized infrared light is used, the following ratio of intensities can be assessed:

$$\log(I_a/I_o)/\log(I_r/I_o) = \beta_a/\beta_r = \phi \equiv \text{dichroic ratio}$$

where a and r represent axial and radial polarization. Infrared light excites specific vibrations in organic molecules. Hence, infrared absorption provides information on molecular structure. Chemists use infrared absorption to characterize organic materials and proof structures. The dichroic ratio provides information on the orientation of the specific groups of the molecules responsible for the absorption. Thus, if an absorption associated with the stretching of, say, a C—C bond is analyzed using the dichroic ratio, we can determine what portion of the C—C segment is oriented along the fiber axis and what portion is oriented normal to the fiber axis. The results complement those of X-ray diffraction and birefringence, but give more specific information on particular molecular groups. An example of the sort of data that can be obtained using dichroic ratios based on IR data is shown in Figure 12.12. In this plot we see the effect of draw ratio on the dichroic ratio of four infrared active absorbances in polyethylene. The absorbances at 1368, 1303, and 1078 cm^{-1}

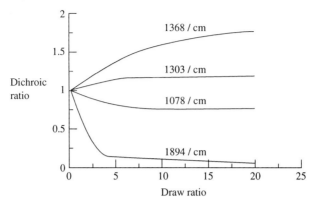

FIGURE 12.12 Effect of draw ratio on dichroic ratio of various infrared bands in polyethylene drawn at 60 °C. (W. Glenz and A. Peterlin, *Journal of Macromolecular Science, Physics* B4 (1970), 473.)

wavenumbers originate entirely from different groups in the amorphous phases, whereas that at 1894 cm^{-1} is due to groups in the crystalline regions. The data show the crystals are highly oriented and approach maximum orientation at low draw ratio.

Dichroic ratios of films is routinely calculated using an infrared absorption unit equipped with polarizers and data processing capabilities, typically Fourier transform IR units. It is, however, *not* a simple matter to measure quantitative data on fibers. Fibers have highly curved surfaces, leading to a great deal of scattering from the surfaces. The fraction of incident radiation reflected from a curved surface depends on the wavelength of the radiation. Squishing devices that apply pressure to flatten fiber surfaces have been introduced, but any induced plastic deformation may lead to molecular rearrangement. Refractive index immersion fluids that do not absorb in the regions of interest have also been used with limited success. Beam condensers that produce a beam width of a few microns have recently been developed, giving rise to the IR microscope.

We have mentioned only a few techniques for assessing molecular orientation—X-ray diffraction, IR dichroism, and birefringence—from the entire repertoire of available ones. Other techniques include fluorescence polarization, solid-state nuclear magnetic resonance, and Raman spectroscopy (Ward, 1975.)

12.5 Summary

The optical properties of materials, like magnetic and electronic properties, are related to the polarizability of the atoms and molecules present in the material. In solid polymers the chemistry of the macromolecules determines the polarizability. Since polymers can be oriented, polymers may have different refractive indices in different directions. In fibers we are often interested in the birefringence, which is the difference between the refractive index parallel to the fiber axis and that normal to the fiber axis. Birefringence is used to assess molecular orientation in anisotropic polymers, especially fibers. The birefringence of fibers with different compositions cannot be usefully compared, because fiber chemistry determines the maximum achievable, or intrinsic, birefringence. Most fibers have positive birefringence.

The surface of fibers reflects a portion of incident light. The greater the relative exposed surface area, the greater the amount of reflection. The higher the refractive index, the greater the reflectivity. Internal surfaces and heterogeneities also reflect light. A material characterized by a high amount of diffuse scattering appears white. A surface that is characterized by a high amount of spectral reflection is a mirror.

Materials that absorb or reflect light preferentially by wavelength in the visible region of the spectrum are colored. The greater the absorption in a translucent or transparent material, the deeper the color. We can color fibers by adding pigments, which are colored particles, or dyes, which are chemicals that bond to the structure of the polymer. Dichroic ratios may be used to assess the molecular orientation of a polymer, but meaningful quantitative IR measurements on fibers are difficult to obtain.

References

D. I. Bower and W. F. Maddens. *The Vibrational Spectroscopy of Polymers*. Cambridge, MA: Cambridge University Press, 1989.

P. R. Griffiths and J. A. deHaseth. *FTIR Spectroscopy*. New York: J. Wiley, 1986.

W. D. Kingery, H. K. Bowen, and D. R. Uhlmann. *Introduction to Ceramics*, 2nd ed. New York: Wiley-Interscience, 1976.

R. Meredith, *Journal of the Textile Institute* 37 (1946), T205.

W. E. Morton and J. W. S. Hearle. *Physical Properties of Textile Fibers*, 3rd ed. Manchester, England: The Textile Institute, 1993.

J. P. Schaffer, A. Saxena, S. D. Antolovich, S. B. Warner, and T. H. Sanders, *Materials Engineering*. New York: Times Mirror Books, 1995.

E. E. Wahlstrom. *Optical Crystallography*, 5th ed. New York: J. Wiley and Sons, 1979.

I. M. Ward, ed. *Structure and Properties of Oriented Polymers*. London: Applied Science Publishers, 1975.

Problems

(1) The birefringence of unoriented PAN is 0.0 and that of highly oriented PAN is 0.0. How can this be so?

(2) Explain qualitatively why a green glass is green and a blade of grass is green. (Note that the former is transparent and the latter is opaque.)

(3) What is the physical basis of Gladstone and Dale's law?

(4) The hottest area in textiles in the 1990's is microfibers. They are fibers with a denier less than 1.0. What important attributes might microfibers give fabrics made therefrom? What might the drawbacks be? Consider how light must interact with a fiber to give it a deep color.

(5) An oriented fiber has a refractive index of 1.56 normal to and 1.66 along the fiber axis. Calculate the refractive index of an unoriented fiber with the same chemistry.

(6) How does Δn change along the $\sigma-\epsilon$ curve of PET?

(7) Why is Δn for PPTA (Kevlar®) in Table 12.1 so high?

(8) What information does the dichroic ratio provide?

(9) What characteristics are desirable for a delusterant in PET?

(10) Show how a fiber in air illuminated using light with parallel rays from below focuses the light. If you viewed a fiber through a half cylinder of glass, what would you see?

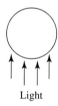

Light

(11) Express the birefringence of a fiber in terms of
   (a) light velocities.
   (b) retardation.
   (c) the angles of incidence and refraction.

(12) Give several techniques that may be used to increase the amount of spectral reflection from a fiber. How can the spectral reflection of a fiber be increased to enhance images in reflected light microscopy? How can spectral reflection be increased to 100% for fibers used in space to prevent penetration and degradation by UV light?

(13) Discuss the roles of specular and diffuse reflection in fiber luster.

(14) Textile grade PET fiber has a birefringence of 0.170, and that of textile grade PP fiber is only 0.030. Provide a fundamental reason for the difference.

(15) How does suntan oil keep you from getting a sunburn? Why do body builders apply oil just prior to competition?

(16) Examine the structure of Vectra® and estimate a reasonable value for the fiber birefringence.

(17) Given that mature cotton fibers appear light blue–green–yellow under crossed polars, explain why immature fibers appear differently, as purple–deep blue.

(18) Why might n_a^o and n_c^o differ? (Hint: Recall Gladstone and Dale.)

(19) Return to the points taught by problems 12 and 15. Explain why refractive index fluids are used in transmitted light microscopy of fibers.

(20) Consider a fiber that has a pale color in transmitted light and is placed in a special reflecting cavity. The arrangement of fiber in the cavity and the fiber surface geometry allow only yellow–orange light to be reflected off the surface. White light is directed into the reflecting cavity. What color will the fiber appear? Why? (This is the origin of the vivid coloring of some insects.)

13

Thermal Properties

The thermal properties of fibers are discussed in this chapter. Similar to the presentations in the last chapter, we will entertain concepts regarding both the thermal properties of fibers and the thermal behavior of fibers. Depending on the application, thermal properties may be of little importance or, on the other hand, critically important. Consider, for example, the significance of the thermal conductivity of the fibrous insulation in cold-weather protective gear to the people exposed to Arctic environments for prolonged periods of time. We will in turn address a number of thermal properties: heat capacity, thermal conductivity, melting and the glass transition temperature, degradation, heat setting, and the coefficient of thermal expansion.

13.1 Heat Capacity

The heat capacity at constant pressure is the amount of energy required to raise the temperature of a unit mass of material one degree. Specific heat is heat capacity divided by molecular weight. The units of specific heat we use are J/kg-K. Specific heat can be viewed as the ability of a material to store energy in the form of heat. In general, organic materials have high specific heats because it is necessary to excite vibrational, rotational, and translational modes of molecular motion in order to raise the temperature. Partially because of their high specific heat, polymer melts can give serious burns when brought into contact with skin. The specific heat at constant pressure, C_p, of a

TABLE 13.1   Thermal Properties of Materials

Material	Density (g/cm^3)	Specific Heat (J/kg-K)	Thermal Conductivity (W/m-K)
Aluminum	2.7	903	237
Copper	8.9	385	401
Steel	7.85	434	60.5
Alumina	4.0	765	46
Boron	2.5	1105	27.6
Carbon fiber	1.8–2.1	710	15–500
Polytetrafluoroethylene	2.2	938	0.35
PET	1.37	1103	0.14
Nylon 6, 66	1.14	1419	0.25
PE	0.97	1855	0.24
PP	0.93	1789	0.12
PAN	1.18	1286	
Wool	1.34	1340	
Wool bats*	0.5	500	0.054
Cotton (cellulose)	1.52	1250	0.07
Cotton bats	0.08	1300	0.06
Glycerol	1.26	2427	0.29
Water	1.0	4180	0.6
Snow	0.11	240	0.05
Air	1.16×10^{-3}	1007	2.6×10^{-2}

Sources: R. C. Weast, ed., *Handbook of Chemistry and Physics*, 63rd ed., Boca Raton: CRC, 1982; J. Brandrup and E. H. Immergut, *Polymer Handbook*, 3rd ed., New York: Wiley-Interscience, 1989; *W. E. Morton and J. W. S. Hearle, *Physical Properties of Textile Fibers*, 3rd ed., Manchester, England: The Textile Institute, 1993.

number of materials is provided in Table 13.1. In general, polymers have larger specific heats than other solids and less than those of organic fluids. Polymers with the greatest rotational freedom have the largest values. A typical organic fiber has a specific heat of about 1300 J/kg-K.

13.2 Thermal Conductivity

Thermal conductivity is the proportionality constant between heat flux and temperature gradient, as defined by Fourier's law, which was discussed in Chapter 6:

$$\text{flux} = q = (\text{flow/area})/\text{time} = -k(dT/dx)$$

where

dT/dx = the temperature gradient, and

k = thermal conductivity.

For a plane wall maintained with constant surface temperatures:

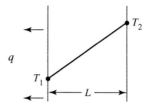

$$q = k(\Delta T/L) = k(T_2 - T_1)/L$$

where
T_1 = temperature of the cool wall,
T_2 = temperature of the hot wall, and
L = wall thickness.

Values of thermal conductivity for various materials are given in Table 13.1. These values are meant to be guidelines only, since molecular orientation can affect thermal conductivity. For example, the thermal conductivity of highly oriented gel-spun polyethylene fiber may be as high as 30 W/m-K (B. Poulaert et al., *Polymer Communications* 31 (1990), 148).

Polymers and textile fibers are thermal insulators. Indeed, textile fibers are often used for thermal protection, such as bats in sleeping bags or winter coats. In these sorts of applications the dead air provides the insulation; hence, a bat that can trap the most air provides the best insulation. If the fiber and air are considered to be in parallel, then the "isostrain," or parallel, model developed in Chapter 11 provides an upper bound calculation:

$$k_{bat} = k_{fiber}V_{fiber} + k_{air}V_{air}$$

where
k = thermal conductivity, and
V = volume fraction.

We design bats so that the second term on the right side of the equation is much less than the first term:

$$k_{bat} \sim k_{fiber}V_{fiber}$$

Since k_{fiber} does not vary by more than about 2 for all textile fibers, the choice of fiber is perhaps best made on the basis of its ability to trap air and minimize V_{fiber} or provide the fabric with resiliency, rather than on the basis of k_{fiber} alone. Similarly, the "isostress," or series, model provides a lower bound calculation:

$$1/k_{bat} = V_{fiber}/k_{fiber} + V_{air}/k_{air}$$

We construct the bats so that the second term on the right dominates:

$$k_{bat} \sim k_{air}/V_{air}$$

Again, we see the need to maximize V_{air}. Thinsulate®, a nonwoven bat manufactured by 3M corporation, for example, contains both (polypropylene) microfibers and coarse, resilient (polyester) fibers. The microfibers provide small convection cells, and the coarse fibers provide resiliency, or the ability to trap a large quantity of air in the bat even after cyclic compression.

The thermal conductivity of a fiber typically increases with moisture regain.

13.3 Melting and Glass Transition Temperatures

13.3.1 Melting

The melting temperature of crystalline solids is a first-order thermodynamic property. At a specific temperature there is a discontinuous change in various properties, for example, volume, as shown in Figure 13.1. Melting is an endothermic process (requires heat). Crystallization is an exothermic process (liberates heat). Under equilibrium conditions, $T_m = T_c$. These facts are well-known principles of equilibrium thermodynamics. Polymers, however, are unusual materials. They are composed of long chain molecules that move slowly and cannot respond to change as quickly as do other, low-molecular-weight materials. Consequently, polymers are usually not in equilibrium unless they are dissolved in large amounts of solvent. Polymers are characterized by a number of unusual properties that are a direct consequence of their long chain nature. Rather than melting at a single well-defined temperature, for example, polymers generally melt over a range of temperatures, well below the equilibrium melting temperature. The reason for the unique behavior is that polymer crystals span a range of size, stability, and perfection. Some are small and imperfect. Others are large and nearly perfect. The observed melting temperature is that at which kT, the thermal input, is sufficient to overcome secondary bonding in the various crystalline regions, and hence, the molecules become fluidlike. Three-dimensional order is lost upon melting. Only crystals melt. Noncrystalline materials do not melt; however, when they are at a sufficiently high temperature that flow occurs, they are referred to as melts. Thermoset or crosslinked polymers do not melt; rather, they degrade.

In Chapter 3 we used the fact that a characteristic amount of heat is absorbed when crystals melt. Heat of fusion, ΔH_f, allowed us to *estimate* a de-

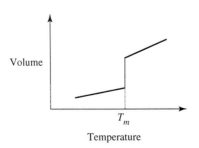

FIGURE 13.1   Change in volume with melting and crystallization.

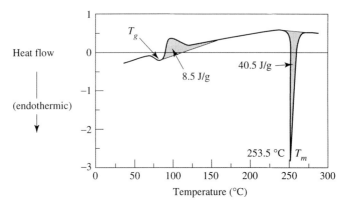

FIGURE 13.2 DSC of poly(ethylene terephthalate) fiber. Note glass transition at 80 °C, crystallization occurring after melting and releasing 8.5 J/g, and melting at 253.5 °C, absorbing 40.5 J/g.

gree of crystallinity, since the amount of heat required to melt a crystal depends on the strength of the secondary bonding in the crystals. The heat of crystallization is the heat liberated when a noncrystalline material crystallizes. The heats of fusion and crystallization can be determined using calorimetry, most readily using differential scanning calorimetry, DSC. In this technique the differential heat required to maintain a sample mass at a temperature equal to that of a reference is plotted as both sample and reference are heated at a fixed rate: T_g appears as a change in slope and T_m as a peak or valley, as shown in Figure 13.2. The heat of fusion (J/g) is the area under the peak normalized to the sample mass. The heat of fusion of pure crystal is generally determined by conducting DSC on standard samples of known crystallinities and extrapolating to 100 percent crystal. Values of ΔH_f (pure crystal) are given in Table 3.1 on page 39.

The absolute value of the melting temperature of a polymer crystal can be understood by considering some elementary thermodynamics. The energy of a material can be expressed as Gibb's free energy, ΔG:

$$\Delta G = \Delta H - T\,\Delta S$$

where

ΔS = entropy change upon melting of the polymer crystals,

T = temperature of the system, and

ΔH = enthalpy change upon melting.

The enthalpy is the net strength of the secondary bonds among molecules in the crystalline regions. The concept of entropy was introduced in our discussion of rubber elasticity. It is a measure of the order of a system. The lower the order, such as polymer molecules coiled upon themselves, the higher the entropy. On the other hand, stiff molecules or sections of molecules—such as aromatic groups—have high order and, hence, low entropy. At the melting temperature, crystal and melt coexist and thus have no difference in energy, so $\Delta G = 0$. Hence,

$$T_m = \Delta H_m / \Delta S_m$$

Melting temperature, therefore, is determined by the quotient of enthalpy and entropy of melting. For polymers it is found that there is a greater range in available entropy difference between crystal and melt than there is for the difference in enthalpy. Hence, unlike in metals or ceramics, the entropy term often dominates in polymers. An example is afforded by liquid crystal polymers. Let us compare the enthalpy of nylon 6 with that of PPTA, poly(paraphenylene terephthamide). Both have hydrogen bonding, so the enthalpy difference among crystals is small. The hydrogen bonding is largely disrupted upon melting, so the enthalpy change upon melting is moderate. On the other hand, PPTA is a rigid-rod polymer with small entropy difference between melt and crystal, whereas nylon 6 is a flexible chain polymer with high entropy difference between crystal and melt. Hence, PPTA will melt at a much higher temperature than nylon 6. In fact, PPTA generally degrades before it melts; however, under appropriate conditions PPTA melts at 500–620 °C. (See Morgan et al., 1983.)

13.3.2 Glass Transition

Polymers generally show a number of transitions in the solid state. These transitions are due to molecular motions that occur in a specific range of temperatures and frequencies of excitation, as described in Chapter 10. An example of a molecular motion that absorbs energy is the crankshaft motion, which is schematically illustrated in Figure 13.3. While these motions are important, we wish to focus our attention on the most important transition below the melting temperature, the glass transition temperature. The glass transition temperature is a property of noncrystalline material only. It is the temperature range over which molecular motion on the size scale of the mer is frozen upon cooling or enabled upon heating. Since a polymer consists of many identical mers, mobility of the entire molecule becomes significant at T_g. Unlike melting, the glass transition is *not* a first-order thermodynamic property; rather, T_g depends on the rate of heating or cooling. A typical volume–temperature curve for an amorphous polymer is shown in Figure 13.4. As a melt is cooled and crystallization is avoided, the V–T curve must bend to avoid crossing the low-

FIGURE 13.3 Schematic of crankshaft motion in a polymer.

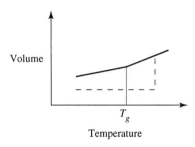

FIGURE 13.4   Change in volume of a material at its glass transition temperature.

est possible energy state, that of the crystal. The temperature of the bend defines T_g.

As discussed in Chapter 3, virtually all polymeric textile fibers, both natural and synthetic, are at least two-phase materials, containing both crystalline and noncrystalline regions. In elastomers some, if not all, noncrystalline polymer is above T_g, since the molecules are fluidlike. In other semicrystalline polymers, the noncrystalline phase may be either above or below the glass temperature.

Two polymers that may be either completely noncrystalline or semicrystalline under ordinary conditions are PET and nylon 66. Given sufficient motivation or encouragement, these two polymers will crystallize—motivation being molecular alignment to reduce the barrier to crystallization, or a considerable amount of time spent cooling between T_g and T_m so as to give the molecules the required time to arrange themselves into crystalline regions, or both. In textile form both polyesters and nylons are semicrystalline, largely because of the high draw imparted to the molecules during processing. Most other synthetic textile polymers, such as PE, PP, PEO, PBT, and nylon 6, crystallize so rapidly that crystallization cannot be easily avoided.

Do not forget, however, that noncrystalline fibers can be formed using ordinary polymers and high cooling rates or by using polymers with such molecular asymmetry that they cannot crystallize. Chemists have intentionally built molecular structures with sufficient asymmetry that molecules cannot pack together, regardless of how much time the molecules have in the fluid state below the thermodynamic melting temperature. Examples are polystyrene, the polycarbonates, and polymethylmethacrylate. These thermoplastic polymers are transparent and may be used in fiber optics or as optical waveguides, but use temperature is limited by T_g. Crosslinked polymers also do not crystallize, because the crosslinks preclude the required molecular motion for crystallization. Examples include epoxies, silicones, and some polyesters.

The equilibrium melting temperature, commonly observed melting temperature, and typical glass transition temperature of various fibers are provided in Table 13.2. The equilibrium melting temperature is the fusion temperature of perfect polymer crystal heated at very slow rates. The equilibrium melting temperature is rarely observed in practice, but it is useful to keep in mind that

TABLE 13.2   Melting and Glass Transition Temperatures of Fibers

Fiber	T_g (°C)	T_m (°C) Equilibrium	T_m (°C) Observed
Wool			none
Cotton	230		none
PET	125	285	255
PE	−80	141	130
PP	−18	187	165
Nylon 66	50	280	260
Nylon 6	90	270	215
PTFE, Teflon®	130	327	no true melting
PAN	90		no true melting
Polybutadiene	11	12	6
Oxide glass	630–785		none

Sources: B. Wunderlich, *Macromolecular Physics: Crystal Melting*, V3, New York: Academic, 1980;
J. Brandrup and E. H. Immergut, *Polymer Handbook*, 3rd ed., New York: Wiley-Interscience, 1989;
S. Kumar, "Advances in high performance fibers," *Indian Journal of Fiber and Textile Research* 16
(1991), 52–64.

it is an upper limit. Please do not misuse these data. Glass transition temperature and observed melting temperature depend on the specific sample used. For example, T_g increases with mer size and molecular orientation, because they make motion in the noncrystalline regions more difficult. These data are provided only to show typical values of thermal properties for various fibers.

13.4   Decomposition and Degradation

Most natural fibers degrade at relatively low temperatures; however, man has need for fibers that can withstand high temperatures. Firemen and jet fighter pilots, for example, need comfortable clothing that maintain their integrity even after exposure to high temperature. While oxide and other ceramic fibers, such as asbestos, glass, or alumina, can be used in some applications, fibers appropriate for use in garments are often needed. Many attempts have been made to stabilize natural fibers (I. Michlik and J. Smekal, *Proceedings of the 4th International Conference on Thermal Analysis* 3 (1974), 425)—imbuing them with fire retardants and stabilizers such as halides and hydrates—but today, perhaps the best approach to *high* thermal stability is to use synthetic fibers with the appropriate chemical structure.

When a cotton fiber is burned, such as was the original preparation by Thomas Edison and Sir Joseph Swan in about 1880 for a filament in an "Ediswan" lamp, the carbonaceous structure formed by controlled pyrolysis, or burning, can withstand high temperatures. Shindo recognized that Edison's observation may be exploited to make carbon fibers. He slowly pyrolyzed an acrylic

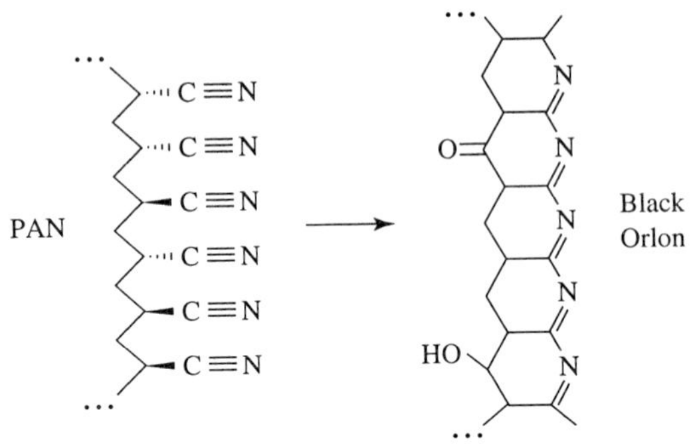

FIGURE 13.5 Schematic structure of black orlon, a ladder polymer.

fiber under tension to develop the first carbon fiber with useful mechanical properties (A. Shindo, *Rpt. 317 Gov. Ind. Res. Inst.*, Osaka, 1961). Shindo's work led to insight in understanding what structural features lead to high-temperature stability.

Chemists now know what sort of structures are required to give polymers high-temperature capability. Several aspects of these fibers' chemical structure need to be addressed:

1. Even the weakest covalent bond in the polymer chain must be strong.
2. The polymer must be clean, as impurities may catalyze decomposition.
3. The polymer needs to be a ladder-type polymer, rather than a single chain polymer, as shown in Figure 13.5.
4. Oxidation of the polymer must be avoided.

Fiber chemistry, then, is the most important consideration in making a high-temperature fiber. Chemists have learned that polyazoles and polyimides are some of the most thermally stable fibers. Polyimides have the basic repeat unit shown:

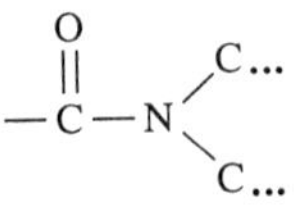

The most thermally stable organic fibers are poly(meta-phenylene isothalamide), or Nomex®; poly(para-phenylene terephthalamide), PPTA, or Kevlar®; poly(benzimidazole), or PBI; and poly(para-phenylene benzobisoxazole), or PBO. Their structures are shown in Figure 13.6. The approximate maximum use temperature of a number of fibers is provided in Table 13.3. Some of these use temperatures are sufficiently high that fibers can be used briefly in open flame, in contact with molten metals, and near a raging fire, especially when the fabrics are aluminized to reflect the bulk of the incident thermal radiation.

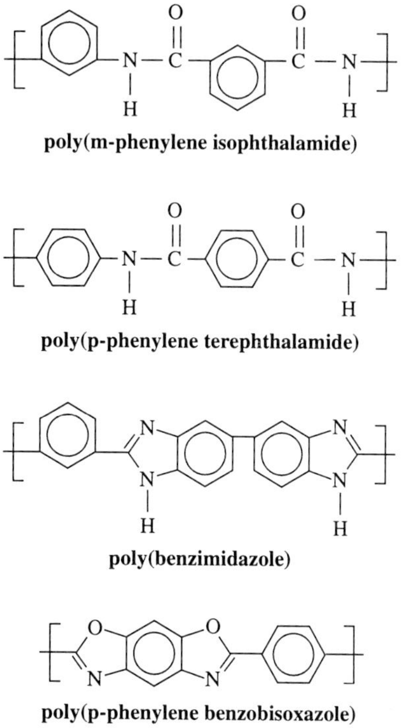

poly(m-phenylene isophthalamide)

poly(p-phenylene terephthalamide)

poly(benzimidazole)

poly(p-phenylene benzobisoxazole)

FIGURE 13.6 Structure of high temperature organic fibers.

TABLE 13.3 **Maximum Use Temperatures of Fibers in Air**

Fiber	Maximum Use Temp (°C)
Cellulosic (cotton)+[a]	150
Animal fiber (wool)+[a]	130
PMIA, Nomex® +[a]	370
PPTA, Kevlar® +[b]	350
PBI+[a]	330
PBO+[b]	600
Carbon~ [b]	500–700
Silicon carbide*[c]	1800
Aluminum oxide+[c]	1540

Maximum use limited by * crystallite melting, + decomposition, ~ oxidation.

Sources: [a] M. L. Joseph, *Introduction to Textile Science*, 5th ed., New York: Holt, Rinehart, and Winston, 1986; [b] S. Kumar, "Advances in high performance fibers," *Indian Journal of Fiber and Textile Research* 16 (1991), 52–64; [c] R. C. Weast, ed. *Handbook of Chemistry and Physics*, 63rd ed., Boca Raton: CRC, 1982.

13.5 Coefficient of Thermal Expansion (CTE)

Most materials expand dimensionally and volumetrically when heated. Polymers obey this general rule. The coefficient of linear expansion, a, is defined by

$$a = (1/l)(dl/dT)$$

where

  l = sample length, and
  dl/dT = slope of the length–temperature curve.

The volumetric expansion coefficient, α, is given by

$$\alpha = (1/V)(dV/dT)$$

where

  V = volume, and
  dV/dT = slope of the volume–temperature curve.

For isotropic materials,

$$\alpha \sim 3a$$

Rather than expanding as anticipated, however, oriented organic fibers typically shrink when heated. Indeed, this is a problem with many garments since most of the shrinkage is irreversible. Thermal shrinkage in fibers is entropically driven. Fiber and other anisotropic polymers have macromolecules extended along the orientation axis. The lowest energy state for fluid polymer chains occurs when they are coiled on themselves. Thus, when provided adequate mobility (kT), the molecules in oriented amorphous or crystalline regions will tend to coil, and hence, fibers and fabrics shrink. Virtually all highly oriented fibers are characterized by $a < 0$, a negative axial thermal expansion coefficient, as shown in Table 13.4. Expansion, of course, occurs in the other two directions. Note that fibers with high helix angles are not characterized by $a < 0$.

13.5.1 Heat Setting

Dimensional stability of a textile fiber in use is often of critical concern. Unpredictable shrinkage during washing of jeans, shirts, or virtually any article of clothing is undesirable. As mentioned, entropy is the driving force for fiber shrinkage and heat, water uptake, or both, can free polymer molecules to move. It is generally the molecules in the noncrystalline regions that, given mobility, seek to coil first; however, it is also possible that small crystallites will melt and the freed molecules seek to coil prior to recrystallization. How, then, can textile fibers be made dimensionally stable? The key is to intentionally expose a fiber in processing to at least its use temperature and allow a small (2 to 3 percent) amount of relaxation to occur, prior to making a fabric from a fiber or yarn. The "set" fiber then becomes dimensionally stable, as shown

TABLE 13.4 Linear Thermal Expansion Coefficients of Polymers

Material	Axial Coefficient of Thermal Expansion (10,000/°C)
Nylon	1
Nylon fiber	−3
PET	0.2
PET fiber	−10
PE	2
PE fiber, Spectra® 1000 (uniaxial composites)	−0.1
PPT, Kevlar® 49	−0.04
Cotton fiber	4
PAN	2
Acrylic fiber	10
Oxide e-glass fiber	0.05
T-300 carbon fiber	−1.2
Aluminum oxide fiber	0.07
Silicon carbide fiber	0.03

Sources: R. C. Weast, ed. *Handbook of Chemistry and Physics*, 63rd ed., Boca Raton: CRC, 1982; J. Brandrup and E. H. Immergut, *Polymer Handbook*, 3rd ed., New York: Wiley-Interscience, 1989; S. Kumar, "Advances in high performance fibers," *Indian Journal of Fiber and Textile Research* 16 (1991), 52–64.

in Figure 13.7 for a PET fiber. The fiber is allowed to shrink a few percent in the relaxation step. Heat is provided by a hot roll, a hot shoe, steam, hot water (which may plasticize the glassy regions and lower T_g), or whatever. A small portion of the orientation that you have worked hard to impart is lost, chiefly in the noncrystalline regions; however, it is generally necessary to use this step to obtain thermal stability in semicrystalline fibers. Relax is not required in fibers that are continuous crystals (Spectra® or Kevlar®) or in applications where use temperature will not be at elevated temperatures.

Textile fiber dimensional stability is commonly measured by boiling water shrinkage, which is the amount of shrinkage that occurs when a fiber is placed in boiling water:

$$BWS = -(l - l_o)/l_o$$

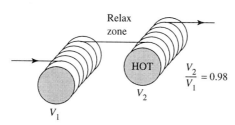

FIGURE 13.7 Heat setting achieved at 120 °C with 2% relax.

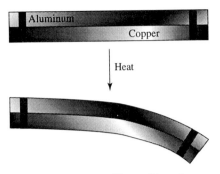

FIGURE 13.8 Bimetallic strip.

where

l = length of the fiber, and

l_o = initial length.

The experiment is simple, the fiber rapidly comes to equilibrium, the water is a constant temperature bath, and the test approximates a severe washing. Tension is generally not applied during the test, but may be used when the water-induced shrinkage is intended to allow some relaxation in samples to be used for further testing.

13.5.2 Thermal Expansion Mismatch

When composites are made using fibers and matrix, residual stresses can develop during cooling from formation temperatures or during use. The difference in thermal expansion coefficient (CTE) between fiber and matrix causes residual stresses. When the difference is zero, no stresses develop. When the differences are sufficiently great, the composite may warp or outright fail. Consider two examples, both circuit boards. In one case, the circuit board is a ceramic fiber–ceramic matrix composite. In the second case, the circuit board is formed by bonding organic fiber with thermoset polymer. When the fibers are not symmetrically positioned in the polymer composite, warping may occur during cooldown in fabrication or in use. This sort of defect or failure is common. The warping is similar to that of a bimetallic strip, as shown in Figure 13.8, or to a sheet of plywood in which one face is exposed to damp conditions. Ceramic fiber–ceramic matrix composites are typically formed at very high temperatures. Differences in thermal expansion between fiber and matrix may lead to high residual stresses in the composite at room temperature regardless of fiber positioning. The stresses may be sufficiently high that the brittle ceramic matrix fails from the tension that is brought about by fibers that shrink less axially than matrix during cooling from fabrication temperature.

The stresses that develop because of thermal expansion mismatch can be estimated using the equation

$$\sigma = E \Delta a \, \Delta T$$

where

E = modulus of material,

Δa = thermal expansion difference between fiber and matrix, and

ΔT = temperature difference between formation and use temperature.

The sign of the stress is easily deduced by inspection. For example, when a fiber expands more than a matrix, the fiber goes into compression and the matrix goes into tension. Compressive stresses are usually easily borne by ceramic elements, but ceramic matrices are especially poor in tension. Low or negative CTE fibers can be used to make composites, especially ductile metal matrix composites, with near zero CTE, for example, C fiber in Al.

13.6 Summary

This chapter has focused on the thermal properties and behavior of fibers. Thermal conductivity is an important property in certain applications, such as in thermal insulation; however, the thermal conductivity of most organic fibers does not depend strongly on chemical composition or microstructure. When used in high-loft applications designed to provide thermal insulation, a fibrous assembly's ability to trap air is of paramount importance. The thermal expansion coefficient of most unoriented polymers is higher than that of other materials. Increasing the molecular orientation reduces the expansion coefficient along the direction of orientation. Highly oriented fibers have negative thermal expansion coefficients. The specific heat of polymers is greater than that of most other condensed phases.

The most highly thermally stable polymers are ladder polymers, especially polyimides. With appropriate molecular architecture, the maximum use temperature of polymer fibers can exceed 600 °C.

Melting is a first-order thermodynamic property of crystalline materials. Polymers do not usually show melting at their equilibrium melting temperature; rather, the observed melting temperature is below the equilibrium melting temperature, largely because crystals are small and imperfect. The secondary bonds, which are maximized in the crystalline regions, are overcome by thermal energy at the melting temperature. The melting temperature is determined by the quotient of the enthalpy of melting and the entropy of melting. In polymeric systems the entropy term often dominates the melting behavior.

The glass transition temperature is a property of noncrystalline materials. It is the temperature range over which mers in the noncrystalline regions become mobile upon heating or immobile upon cooling. Other transitions are due to motions in the polymer that are on a scale smaller than a mer. A semicrystalline polymer will retain a substantial fraction of its mechanical properties up to the melting temperature only if the crystallinity is reasonably high. Otherwise, properties degrade at the glass temperature. Heat setting may be regarded as a thermal–mechanical process that allows for a minor amount of shrinkage in the noncrystalline regions and perhaps some recrystallization. A small portion of molecular orientation is sacrificed for dimensional stability.

References

J. Brandrup and E. H. Immergut, eds. *Polymer Handbook*, 3rd ed. New York: Wiley-Interscience, 1989.

K. K. Chawla. *Composite Materials*. New York: Springer-Verlag, 1987.

M. L. Joseph. *Introductory Textile Science*, 5th ed. New York: Holt, Rinehart, and Winston, 1986.

S. Kumar. "Advances in high performance fibers." *Indian Journal of Fiber and Textile Research* 16 (1991), 52–64.

R. J. Morgan, C. O. Pruneda, and W. J. Steele, *Journal of Polymer Science; Polymer Physics Edition* 21 (1983), 1757.

W. E. Morton and J. W. S. Hearle. *Physical Properties of Textile Fibers*, 3rd ed. Manchester, England: The Textile Institute, 1993.

R. C. Weast, ed. *Handbook of Chemistry and Physics*, 63rd ed. Boca Raton: CRC Press, 1982.

Problems

(1) The following observations are made on a polymer:

(i) It is quite stretchy and recovers completely at room temperature.

(ii) Below room temperature, it becomes opaque and stiff.

(iii) At cryogenic temperatures, it becomes much stiffer and brittle.

Briefly describe a structure of the material that is consistent with the observations.

(2) A sample of a new polymer is made into an *oriented fiber* simply by dipping a glass rod into the polymer melt and drawing the polymer upward. X-ray analysis shows that the fiber is 30% crystalline and that the crystals are cylindrical—about 100 Å in diameter and 1 μm long. The density of the fiber is 1.25 and that of the crystal alone (by X-ray diffraction) is 1.44 g/cm^3. Pure crystal melts at 200 °C with a heat of fusion of 180 J/g.

(a) Sketch what you anticipate the DSC will show for the material, including a value for the heat of fusion.

(b) Calculate the density of the noncrystalline material.

(c) What model might you use to estimate the modulus, given values for crystalline and noncrystalline regions? Describe how the composite responds to tensile deformation.

(d) Sketch the effect of temperature on modulus of the material.

(3) The CTE (thermal expansion coefficient) of an unoriented sample of polyester is 2×10^{-5}/K. That of an oriented fiber is 6×10^{-6}/K along the fiber axis. Realizing that CTE is an intrinsic property, estimate CTE of the oriented fiber normal to the fiber axis. Be sure to let me know what an intrinsic property is.

(4) PET is a thermoplastic polymer and cotton is a thermoset polymer. Explain the following observations:

(a) PET has a T_g of 110 °C and that of cotton is 225 °C.

(b) PET melts at 255 °C and cotton degrades prior to melting.

(c) PET degrades at about 400 °C and cotton degrades at 245 °C.

(5) A 1 kg brick of PE is heated to 70 °C. The brick is placed in an insulated bath containing 1 kg of water at 10 °C. No water escapes. No heat leaves the system. At what temperature does the system equilibrate? Repeat for a 1 kg mass of steel.

(6) Suppose you are in the kitchen making candy. This candy requires steady heat to impart the appropriate viscosity. The recipe requires you to add baking soda, stir it in, and wait for the baking soda to react with the vinegar, forming bubbles. The candy is quite viscous at this time, so the bubbles create a foam. This is what you want; however, the candy quickly burns (carbonizes) soon after the foaming began. You repeat the procedure, being certain that you do not change the burner setting, but the candy burns again. Why does this occur? How do you prevent burning?

(7) Show why molecular orientation increases the glass transition temperature.

(8) Explain why promoting further crystallization in an oriented polymer reduces the orientation of the molecules in the noncrystalline regions and thus reduces the potential for shrinkage with temperature.

(9) Show that $\alpha \approx 3a$.

(10) Several years ago, Eastman Chemical introduced a new polymer called X7G. It was a copolymer based on PET, so the anticipated use temperature was roughly 250 °C. DSC confirmed that crystals melted at a few degrees higher than 250 °C. Softening tests showed that the material could not be considered a solid above 125 °C. What is the structural reason for the low maximum use temperature?

(11) Use elementary thermodynamics to explain why PET has a higher melting temperature than any of the wholly aliphatic polyesters.

(12) Suggest why the use temperature of a carbon fiber may be as low as 500 °C although carbon does not melt or sublime below 2800 °C.

(13) Discuss some of the pros and cons associated with using polyester bat rather than goose down as sleeping-bag insulation. What gives the goose down both resilience and good insulation?

14

Fiber Friction, Electrical Conductivity, and Static Charge Effects

Fiber rubbing and friction may lead to equipment failure and surface damage in fibers. Surface damage, in turn, reduces fiber strength. On the other hand, without friction among fibers, staple fibers cannot be processed into yarns and yarns have no strength. Fiber rubbing also contributes to static charge buildup. Static charge does not rapidly dissipate because organic fibers are electrical insulators. Clearly, the three topics of this chapter—fiber friction, electrical conductivity, and static charging—are related. We will address the fundamental concepts associated with each one in a logical order. There has been a great deal written about each of these topics, generally in the form of empirical studies. Rather than attempting to cover all the details, what I shall do is present the basic principles associated with each topic and give a few examples or problems.

14.1 Fiber Friction

Fiber friction is something you "can't live with and can't live without." It has a dualistic role in fiber technology.

Fiber Friction in Fiber Technology

High Friction Required	Low Friction Required
1. Enable drafting	1. Yarns passing through guides
2. Provide effective transfer of fiber strength to yarn strength	2. Minimize wear of fibers and fabrics
	3. Provide good fabric drape

Friction, fiber-on-fiber or fiber-on-other-solids, can lead to surface damage, weak fibers, and even fiber breakage. On the other hand, friction is the key force that holds staple fiber yarns together. Without friction, yarns cannot be drafted. Without friction, yarns have no strength.

Just what is friction? The coefficient of friction, μ, is defined by the equation:

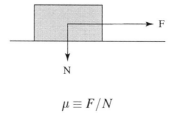

where
$$\mu \equiv F/N$$

F = force to move one surface over another, and
N = normal force pressing the surfaces together.

By 1699, Amontons had presented the basic laws of friction:

1. Frictional forces are independent of the areas of contact.

2. Frictional force is proportional to the normal force between surfaces in contact.

In addition, Amontons established the concept of static and kinetic friction, realizing that static friction, the force that must be overcome to begin sliding, was always at least as great as kinetic friction, the force that must be overcome to continue sliding. For many materials kinetic friction is independent of sliding speed; however, the behavior is complex in semicrystalline polymers. They show no simple functional dependence of the coefficient of friction on sliding speed, as shown in Figure 14.1.

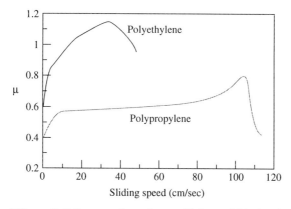

FIGURE 14.1 Effect of sliding speed on the coefficient of friction in semicrystalline polymers. (K. G. McLaren and D. Tabor, *Nature* 197 (1963), 856–858, MacMillan Magazines Limited. Reprinted with permission from *Nature*.)

The Capstan equation can be used to calculate the development of tension as a fiber or yarn passes over a cylindrical surface:

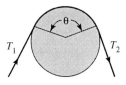

$$T_2/T_1 = \exp(\mu\theta)$$

where

T_1 = incoming tension,

T_2 = outgoing tension, and

θ = angle of wrap or contact.

Calculations based on Amonton's law show how quickly tension develops as a fiber or yarn passes over rolls. The results are presented in Table 14.1. Unfortunately, quantitative analysis shows that Amonton's laws are not obeyed: The coefficient of friction is a function of the normal load, as shown schematically in Figure 14.2. The simplest equation to describe the relationship between frictional force, F, and normal force, N, is the empirical one:

$$F = aN^n$$

where

a and n are constants that depend on fiber and direction of rubbing, and

n generally falls between 3/4 and 1; 0.9 may be used in the absence of a better value.

The study of friction and its effects is largely experimental. Tables 14.2 and 14.3 are designed to provide a general feel for typical values of fiber friction. They also give values of fiber friction that may be of use in a variety of problems. The tables do not suggest the consequences of friction. Fiber friction leads to surface damage, fiber deformation, and fiber breakage, the extent depending in part on the type of friction encountered: hysteresis, plowing,

TABLE 14.1 Tension Development on Cylindrical Rolls According to Amontons

	T_2/T_1		
μ	$\theta = \pi/2$	$\theta = 2\pi$	$\theta = 4\pi$
0.2	1.4	3.5	12.3
0.5	2.2	22.9	525
0.9	4.1	285	81,146

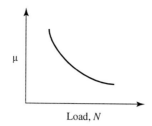

FIGURE 14.2 Effect of normal load on coefficient of friction.

rolling, abrasion, etc. Fiber scientists have learned to use lubricants as appropriate and to avoid lubricants at other times. In this way they are able in part to control surface friction and static charge development. Another important effect of friction that the tables do not convey is the damage that fibers can do to machinery. As described in the last chapter, synthetic fibers often contain small additions (0.1 to 0.5 percent) of titania. TiO_2 is highly abrasive and will wear metal, and to a lesser extent, ceramic parts such as guides and slip rolls. Titania can destroy a metal wear surface in just a few weeks. A scanning

TABLE 14.2   Fiber-on-Fiber Friction

	μ	
Fibers	Static	Kinetic
Rayon-on-rayon[a]	0.35	0.26
Nylon-on-nylon[a]	0.47	0.40
Wool-on-wool[b]		
with scales	0.13	0.11
against scales	0.61	0.38

Sources: [a]B. Olofsson and N. Gralen, *Journal of the Textile Institute* 43 (1950), 467; [b]W. Zurek and I. Frydrych, *Textile Research Journal* 63 (1993), 322.

TABLE 14.3   Friction of Fibers on Solids

	μ			
	Fiber (axial)[a]	Fiber (normal)[b]	Steel[c]	Porcelain[c]
Nylon	0.14–0.6	0.47	0.32	0.43
Rayon	0.19	0.43	0.39	0.43
Cotton	0.29,0.57	0.22	0.29	0.32
Polyester		0.58		
Oxide glass		0.13		
Polytetrafluoroethylene		0.04		

Sources: [a]B. Olofsson and N. Gralen, *Journal of the Textile Institute* 43 (1950), 467; [b]E. H. Mercer and K. R. Makinson, *Journal of the Textile Institute* 38 (1947), T227; [c]H. Buckle and J. Pollitt, *Journal of the Textile Institute* 39 (1948), T199.

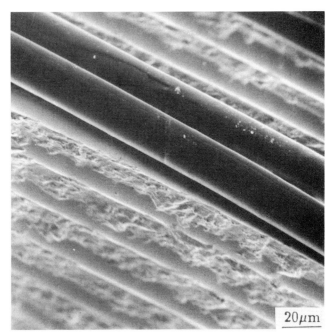

20μm

FIGURE 14.3 Grooves in the surface of a titania guide caused by fiber friction and wear. (P. M. Ramsey and T. F. Page, *Textile Research Journal* 62 (1992), 715, Fig. 1.)

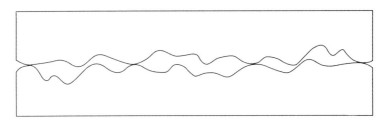

FIGURE 14.4 Schematic of surface contact.

electron micrograph of the sort of damage that is associated with fiber–guide friction is shown in Figure 14.3. Maintenance of textile machinery is a significant cost in a textile operation.

14.1.1 Theoretical Aspects of Friction

What is the fundamental cause of friction? On a microscale virtually all surfaces are rough, as shown in Figure 14.4. Surfaces touch at the high points, or asperities, where considerable compressive and shear stresses lead to plastic deformation. The energy associated with plastic deformation causes localized heating and diffusion, and, hence, material on the contact points tends to

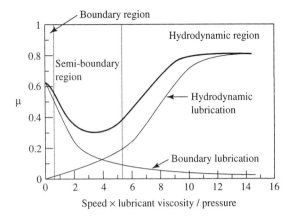

FIGURE 14.5 Lubricated friction. The plot in Figure 14.5 is not set up in a standard fashion: note that the boundary layer region extends to a fiber speed of about 0.1 m/min and the hydrodynamic region begins at about 5.5 m/min.

weld. Fresh, clean surfaces are not poisoned with dirt. Atoms or molecules can diffuse across clean surfaces in contact, leading to primary bonds bridging the interface. The process is accelerated by the heat developed with plastic deformation. (Recall that plastic deformation is not recoverable, and hence the work goes chiefly into heat. Also, polymers have poor thermal conductivity.) Lubrication separates surfaces and minimizes contact, heat generation, mass transfer, and wear.

The total frictional force is the sum of two terms, the force of adhesion and the force of deformation. As just described, the force of adhesion is a result of polymer sticking. Relative temperature is important to adhesion. Because of its very low surface energy, polytetrafluoroethylene does not show adhesion. This gives PTFE the lowest coefficient of friction of any material ($\mu = 0.04$). The deformation force is a result of plowing, to which a polymer responds with hysteresis losses. Surface roughness is important in that it determines the extent of hysteresis loss.

14.1.2 Lubrication

Most surfaces subjected to unwanted friction are lubricated. A good lubricant for polymers is a material that is characterized by a low shear strength and that adheres strongly to the polymer surface. At high processing speeds the amount of heat generated can lead to extremely high temperatures. Consequently, lubricants need to have good thermal stability. When a lubricant is applied, there are three zones of lubrication established, their position depending largely on processing speed: boundary, semi-boundary, and hydrodynamic, as shown in Figure 14.5. Reductions in friction are achieved by imposing a low shear strength fluid in the boundary region. Lubrication tends to be intermittent. (In the absence of a lubricant, the polymer itself shears.) Stick–slip phenomena, which are characteristic of all materials, generally occur in the boundary region.

Hydrodynamic friction is achieved by imposing a continuous layer of lubricant between sliding surfaces. Surface roughness (either metal or fiber) reduces friction in the hydrodynamic region, but not in the boundary region. Even though high friction may exist in the hydrodynamic region, wear rates in the boundary region can exceed those of the hydrodynamic region. For example, in one study of dull PET in contact with a chrome pin, the wear rate in the hydrodynamic region was 9 μg/100 g of yarn with $\mu = 0.7$. In the boundary region, $\mu = 0.25$, but the wear rate was 80 μg/100 g yarn (J. S. Olsen, *Textile Research Journal* 39 (1969), 31). These data further support the hypothesis that lubrication is intermittent in the boundary region, which may or may not be caused by incomplete coverage by the lubricant or finish. (See Kamath et al., 1993.)

14.2 Electrical Conductivity

Like most polymers, textile fibers are electrical insulators. This is due to the fact that all the electrons are bound to the nuclear cores or shared in the covalent bonds. No electrons are free to move. Perhaps surprisingly, the conductivity of some textile fibers changes ten orders of magnitude with relative humidity, as shown in Figure 14.6. Let us explore the reason for this behavior. Textile fibers are insulators or large band gap materials. There are no free electrons in textile fibers, and a large thermal activation, kT, is required to promote electrons from the top of the valence band to the bottom of the conduction band, as shown in Figure 14.7. This behavior may be contrasted to that of metals, where there is no gap and, hence, many free electrons at room temperature. Semiconductors are materials with small band gaps. In

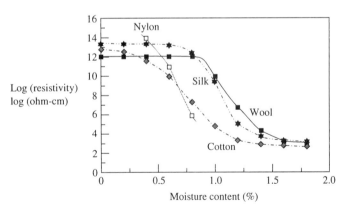

FIGURE 14.6 Electrical conductivity of fibers. (Data from J. W. S. Hearle, *Journal of the Textile Institute* 44 (1953), T117; and G. E. Cusick and J. W. S. Hearle, 46 (1955), T699.)

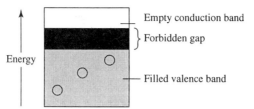

FIGURE 14.7   Band diagram of textile fibers.

these materials kT at room temperature is sufficient to promote a number of electrons from the valence band through the gap into the conduction band. Consequently, semiconductors have conductivities between those of conductors and insulators, and the conductivity is a strong function of temperature.

The basic equation that governs electrical conductivity, σ, is

$$\sigma = nq\mu$$

where

n = number of charge carriers per unit volume,
q = charge per carrier, and
μ = mobility of the carrier.

This equation simply states that conductivity is due to a concerted motion of charge carriers.

Conductivity is the reciprocal of resistivity, ρ:

$$\sigma = 1/\rho$$

Resistivity is a geometry-independent material property that is related to resistance, R:

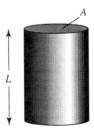

$$\rho = R(A/L)$$

Since there are no mobile electrons in textile fibers, ionic species are responsible for the low but finite electrical conductivity. The ionic species are chiefly impurities, residues from nature in cotton or wool, or catalyst fragments from organic synthesis, etc. When fibers are purified or intentionally doped with salts, the conductivity decreases or increases in a manner consistent with these

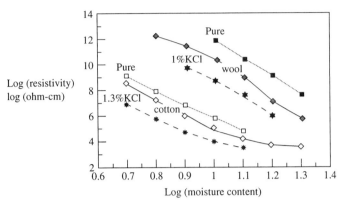

FIGURE 14.8 Effect of ion concentration modification on electrical conductivity of fibers. (J. W. S. Hearle, *Journal of the Textile Institute* 44 (1953), T117.)

arguments, as shown in Figure 14.8. A look back at Figure 14.6 shows that it is the hygroscopic fibers, the ones that absorb water, that show the largest changes in conductivity with moisture content. Both the mobility of the ions and the number of charge carriers increase with moisture content. Water enhances diffusion rate and, more importantly, the disassociation of charged complexes. The net result is a modest increase in mobility and a dramatic increase in the number of charge carriers with RH. Hydrophilic fibers such as PE and PP are generally good insulators, have negligible charge mobility, and show little or no moisture sensitivity. In fact, highly charged solutions or melts of PE or PP can be spun directly into fibers using electrostatic forces alone (L. Larondo and R. St. John Manley, *Journal of Polymer Science: Polymer Physics Edition* 19 (1981), 909, and D.-I. A. Weghmann, *Nonwovens Industries*, (Nov. 1982), 24).

Conductivity in hygroscopic textile fibers is largely by motion of positively charged impurity ions through the bulk structure. Obviously, the noncrystalline regions are more amenable to both moisture penetration and motion of ions than are the more dense and ordered crystalline regions.

In some cases a fiber will conduct charge along its surface, especially when the surface is wet with adsorbed moisture or treated with a thin conductive layer. This mechanism is generally important in fibers that are wet by water, but do not swell. One technique used to intentionally reduce static charge buildup in fabrics is to coat fibers or yarns with antistats, which are organic compounds that have reasonable conductivity. The enhanced surface conductivity is often sufficient to reduce the static problem. A specific example of an application of this technology is in operating room gowns. Most gowns are treated with a fluorocarbon surfactant that both reduces the contact angle with blood and increases electrical conductivity to a value such that static charge buildup in the operating room theater is avoided (see, for example, R. H. Brock and G. H. Meitner, US Patent 4,041,203 (1979)).

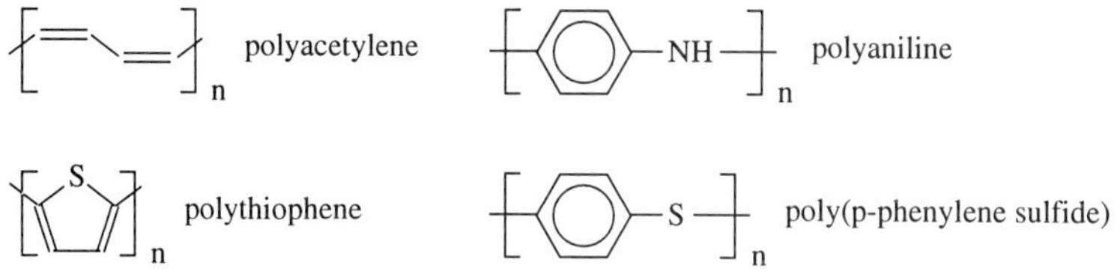

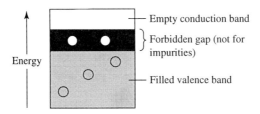

FIGURE 14.9 Chemical structure of organic conductors.

Energy

— Empty conduction band

} Forbidden gap (not for impurities)

— Filled valence band

FIGURE 14.10 The band structure of a polymer doped with an oxidative agent.

14.2.1 Conductive Fibers

Recent advances in polymer chemistry and physics have led to a new class of materials—conductive organic polymers. These materials are characterized by extended pi-bonded, fully conjugated structures, such as fused aromatic rings, double bonds, lone electron pairs, and resonance structures. Examples of conductive polymers are polyacetylene, polyaniline, polythiophene, and poly-(phenylene sulfide), which are shown in Figure 14.9. While the polymers themselves are not conducting, doping with huge amounts of oxidative or reductive reagents such as AsF_5 leads to a change in the band structure. As in the case of extrinsic inorganic semiconductors, the dopant provides available energy levels in the gap—either acceptor (p-type) or donator (n-type) levels. AsF_5, for example, is an oxidant that introduces acceptor levels upon complexation, as shown in Figure 14.10. Electrons from the top of the valence band can be promoted by kT at room temperature to the acceptor level, leaving behind a mobile hole, just as in the case of p-type inorganic semiconductors. The mobile hole leads to enhanced conductivity. As outlined in Table 14.4, specific conductivities (conductivity/density) comparable to that of copper have been achieved. These materials require considerably more research and development before they will be a commercial reality. Many of the polymer complexes, for example, are not stable in air.

14.3 Static Charge Generation and Dissipation

Problems associated with static charging are often considered a nuisance, such as when socks stick to sheets in the dryer; however, the problems can at times

TABLE 14.4   **Conductivity of Doped Polymers**

Material	Conductivity (ohm-cm)$^{-1}$
Poly(para-phenylene)	5×10^2
Poly(para-phenylene sulfide)	5
Poly(acetylene)	10^3
Poly(pyrrole)	10^2
Poly(thiophene)	10
Copper	8×10^5

Source: J. E. Frommer and R. R. Chance, "Electrically conductive polymers" in *Electrical and Electronic Properties of Polymers: A State-of-the-Art Compendium*, J. I. Kroschwitz, ed. New York: Wiley-Interscience, 1989.

result in serious consequences:

• Electronic equipment can be overloaded and break down.

• Parachutes may not open.

• Woven fabrics may "fog mark." The action of a loom charges the yarns and fibers. When the loom is stopped, dirt and dust particles, which are usually charged, will be attracted to the fibers and irreversibly adhere to the yarns, discoloring them.

• Static discharge in an operating room may lead to explosion.

Virtually everyone has experienced the effects of static charge buildup and discharge. How is the charge developed and how can the problem be addressed? Static charging commonly occurs with a number of textile fibers including cotton, wool, rayon, acrylic, and polyester, in a number of textile operations including weaving, opening, drafting, and folding. Static charging requires charge motion, which is impossible in perfect insulators. Perhaps PE, PP, and other hydrophobic fibers do not readily develop static charge because they are such good insulators that few, if any, charges are mobile.

The highest observable charges developed are on the order of 1500 μC/m^2, and these were measured under vacuum. More typical of atmospheric conditions are 30 μC/m^2, which is a concentration of charge that produces an electric field of 3000 kV/m. The mechanisms listed here are theoretically capable of generating values as high as 10^5 μC/m^2. Leakage, generally through the atmosphere, facilitates immediate reductions to 30 μC/m^2.

There are six mechanisms that can lead to static charge development. Any one of these mechanisms is capable of leading to static charging, but in practice, more than one mechanism is usually operative.

1. Suppose a metal is contacted with a polymer. Electrons from the metal cannot pass into the insulator because of the relative positions of the energy levels. Rather, a few levels available to electrons in the insulator within the gap associated with surfaces are occupied. As shown in Figure

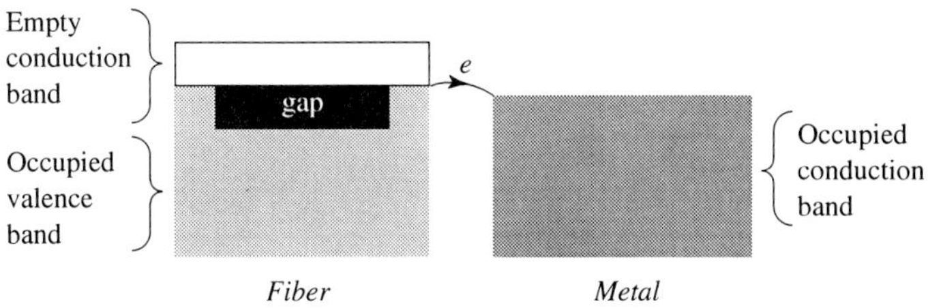

FIGURE 14.11   Band structure of a fiber in contact with a metal.

14.11, the electrons pass into the metal upon contact, giving the insulator a net positive charge, which agrees with experiment.

2. Charged particles on the surface of a fiber redistribute according to Boltzmann statistics. Consider a fiber with acidic groups on the surface, H^+. Upon contact with a different material with a different concentration of H^+ on the surface, ions will move across the surface in numbers that depend on the energy required to transfer, ΔG:

$$n_{jump} = n_o \exp(-\Delta G / kT)$$

where

n_o = pre-exponential constant,

ΔG = activation energy,

k = Boltzmann's constant, and

T = absolute temperature.

This mechanism is important in acidic or basic polymers, or polymers with space charge layers. It is largely responsible for the charging that develops when two different polymers are brought into contact. The electrostatic series developed by Hersh and Montgomery, shown in Table 14.5, cites which polymers tend to develop a positive charge and which develop a negative charge when dissimilar polymers are placed in contact. This is also called the triboelectric series, and it is directly analogous to the galvanic series of metals. Here, however, the charges that develop are relatively small, and there is little current flow.

3. The first two mechanisms involve neither stress nor relative motion. Motion is not required for the development of static charge; however, it can certainly enhance the effect. Asymmetric rubbing produces temperature gradients: Hot spots will develop because of the nature of surfaces and friction. Mobile charges move from hot to cold. Hence, static charge may develop, as sketched in Figure 14.12. Note that this mechanism does not require two different materials.

4. If a material has a charged layer at the surface, then rubbing may skim the surface layer and transfer it to the second material, as illustrated

TABLE 14.5 **Electrostatic Series According to Hersh and Montgomery**

Positive at top, negative at bottom:
Wool
Nylon
Viscose rayon
Cotton
Silk
Acetate
PMMA
Poly(vinyl alcohol)
PET
Acrylics
PE
PTFE

Source: S. P. Hersh and D. J. Montgomery, *Textile Research Journal* 25 (1955), 279.

in Figure 14.13. Recall that one mechanism of friction is adhesion and removal of polymer from the surface.

5. Piezoelectricity is the generation of current (charge motion) upon the application of stress. Wool and other oriented fibers are piezoelectric.

6. Pyroelectricity is the generation of current (charge motion) with increases in temperature. Wool and other fibers are pyroelectric.

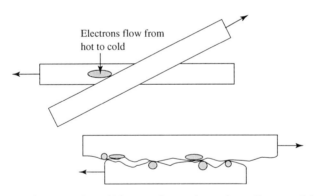

FIGURE 14.12 Asymmetric rubbing produces thermal gradients and both positive and negative charge buildup along a fiber.

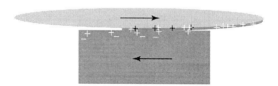

FIGURE 14.13 Skimming of a charged layer.

Static charge development or the decay of static charge must be appropriately dealt with. Textile mills have developed workable solutions, and many fabrics are afforded protection. All electrical and mechanical equipment in textile mills is grounded to protect the equipment and operators. In addition, since humidity increases the conductivity of many fibers, some mills add humidifiers to the central heating and air conditioning system. Spraying a local mist of water or a conductive finish, for example, near the opening, drafting, spinning, or carding areas can also lessen static problems. In finished products antistatic agents are often applied to the fiber surface, providing a rapid conductive path for removal of charge or charge neutralization. In carpets the addition of a conductive fiber, such as carbon or carbon-loaded fiber as shown in Figure 5.8 on page 91, into every yarn reduces the static buildup to acceptable levels.

One application that takes advantage of charges in fibers is filtration. Electrets are polymeric insulators into which charge has been injected: Gasses are ionized by exposure to a high electric field. The ionized gas is accelerated and impacts a polymer, perhaps a fiber. The momentum of the ion drives it into the polymer, giving the fibers a permanent charge. This technology is used to make dust masks with low pressure drops. Dust particles entering a maze of fibers are caught because of their surface characteristics, momentum, charge, or a combination of these factors.

Another use of static charging is in the random deposition of spunbond fibers. Spunbond fibers tend to clump together, giving poor basis weight uniformity. Charged fibers, on the other hand, repel one another, giving rise to fiber separation. By charging spunbond fibers prior to laydown on the conveyor belt, web formation can be improved (see Chapter 15).

14.4 Summary

The essential features of friction between fibers or fiber and solid surfaces have been highlighted. Friction is both a blessing and a curse in textile processes. Fiber processors typically use lubricants to minimize the unwanted effects of friction.

The basic laws of friction as described by Amontons were presented. Unfortunately, Amontons' laws are crude approximations for fibers. Thus, the study of friction and its effects is largely empirical. The coefficient of friction is high for static situations and low for dynamic situations; however, it is not constant with sliding speed. Most surfaces are rough on an atomic scale and the asperities play a large role in friction. Friction has both an adhesive and a hysteresis, or plastic, component. The adhesive component means that polymer is removed from the asperities during sliding contact. The plastic component contributes to the generation of heat.

Lubricated friction has two distinct regions plus a transition region. A good lubricant is one that forms strong secondary bonds with the substrate. In boundary-layer lubrication, the lubricant is a low-shear-strength film between surfaces, providing a layer in which the shear stresses are concentrated. The lubricating action is intermittent. In hydrodynamic lubrication, which occurs at high fiber speeds, the lubricant provides a continuous layer of protection. The friction may be high in the hydrodynamic region, but the wear rate does not correlate with the level of friction.

Most organic fibers are insulators; however, those that absorb water show substantial increases in electrical conductivity with moisture content. The species responsible for the conductivity are positively charged ions that diffuse in a concerted fashion through the water-plasticized noncrystalline regions. Experimental organic fibers capable of reasonable levels of electrical conductivity have been synthesized. Such polymers have extended pi-orbital structures and, in addition, need to be doped or complexed with acceptor or donator impurities to high levels.

Static charge can be developed in fibers simply by contact with another polymer or metal, or by rubbing. Some motion of charge is required for static charge development, so hydrophobic synthetic fibers do not generally pose a static problem. Static problems can be minimized by using conductive finishes such as water or antistats, grounding equipment, adding conductive paths such as conductive fibers in carpets, or increasing the discharge rate by increasing the atmospheric humidity.

References

B. W. Cherry. *Polymer Surfaces*. New York: Cambridge University Press, 1981.

J. E. Frommer and R. R. Chance. "Electrically Conductive Polymers," in *Electrical and Electronic Properties of Polymers: A State-of-the-Art Compendium*. J. I. Kroschwitz, ed. New York: Wiley-Interscience, 1989.

S. P. Hersh in M. J. Schick, ed. *Surface Characteristics of Fibers and Textiles*, Part I. New York: Marcel Dekker, 1975.

H. G. de Jong. *Textile Research Journal* 63 (1993), 14.

Y. K. Kamath, S. B. Ruetsch, and H. D. Weigman. *Textile Research Journal* 63 (1993), 19.

Y. E. El Mogahzy and R. M. Broughton. "New Approach for Evaluating the Frictional Behavior of Cotton Yarns," *Textile Research Journal* 63 (1993), 465; and Y. E. El Mogahzy and B. S. Gupta. *Textile Research Journal* 63 (1993), 219.

W. E. Morton and J. W. S. Hearle. *Physical Properties of Textile Fibers*, 3rd ed. Manchester, England: The Textile Institute, 1993.

J. P. Schaffer, A. Saxena, S. D. Antolovich, S. B. Warner, and T. H. Sanders. *Materials Engineering*. New York: Times Mirror Books, 1995.

M. J. Schick in M. J. Schick, ed. *Surface Characteristics of Fibers and Textiles*, Part I. New York: Marcel Dekker, 1975.

Problems

(1) You are working in a fiber plant processing dull acrylic fiber. Give the pros and cons of using porcelain versus metal guides to direct the yarns.

(2) Is it possible to weld one sheet of polymer to another sheet of the same composition simply by cleaning the surfaces and pressing them together?

(3) Discuss the nature of electrical conductivity in wool.

(4) When you shuffle across the rug and reach to touch a friend, a spark jumps.
   (a) How did the voltage develop?
   (b) What would happen if you waited two minutes before reaching?
   (c) How did the charge get from your socks to your hand?
   (d) If your friend misted you with a fine spray of water, what would occur?

(5) When textile engineers claim "you can't live with and you can't live without friction," what do they mean?

(6) Describe what occurs on a microscopic scale when two textile fibers, say cotton and wool, rub—both electrically and physically (friction).

(7) How does Static Guard® work? It is a product that you spray onto your clothes to keep them from clinging.

(8) Use your knowledge of lubricated friction to discuss the behavior of treaded versus slick automobile tires on dry versus puddled roads.

(9) Show or explain why wool is piezoelectric.

(10) How might you keep a carpet from allowing buildup of static charge on someone shuffling across the floor? a parachute?

(11) Which fiber is more anodic (negative) in a cotton–polyester fabric? Sketch the band structure of cotton in contact with polyester.

(12) Hair conditioner makes just-washed wet hair easy to comb. What is the operative principle of conditioner?

PART FOUR

Fibers to Fabrics

15

From Fibers to Yarns and Fabrics

Our knowledge of fibers has grown considerably in progressing from Chapters 1 through 14. We are indeed more or less experts on fiber structure and properties. We are now interested in how fabric properties depend on properties of the constituent fibers. It seems reasonable to anticipate that the fabric properties depend on a combination of fiber properties and the fabric structure. This expectation is borne out in practice.

We begin our analysis by considering the structure of yarns. A yarn may be envisioned as a collection of nearly parallel fibers, which is usually given some twist. Yarns are made into fabrics, chiefly by interlacing, or weaving, and by interlooping, or knitting. Nonwovens are made directly from fibers, without the yarn-formation steps, but since the fibers are neither interlaced nor interlooped, they must be bonded together. We will see a number of differences in properties between knit, woven, and nonwoven fabrics based on these simple observations. One of the largest differences, and one that we will investigate in considerable detail, is fabric drape. Figure 15.1 summarizes the relationships between fiber, yarn, and fabric properties. The chemistry of a fabric is determined by the fiber chemistry; however, surface chemistry may be altered by finishes applied to the fabric or yarn. The mechanical properties of the fabric are determined largely by the fabric construction and the properties of the yarn, which are in turn a strong function of the fiber properties. We use the term *fabric construction* to refer to the structure of the knit, woven, or nonwoven fabric.

We begin this chapter with a section on yarn structure and properties, both filament and staple yarns. The role of fiber properties in determining yarn

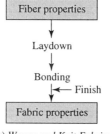

(a) *Woven and Knit Fabrics*

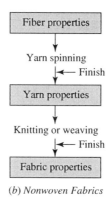

(b) *Nonwoven Fabrics*

FIGURE 15.1   Relationship between fiber, yarn, and fabric properties.

properties is ubiquitous. The next section begins with a brief summary of knit, woven, and nonwoven constructions. The importance of fabric construction and yarn properties to fabric properties is shown and exemplified in a final section on fabric drape. The parameters that influence drape—fabric bending stiffness and fabric shear—are described in knit, woven, and nonwoven structures. We focus on mechanical properties, although other properties could be considered in much the same fashion.

15.1 Yarn Structure and Properties

Before we attempt to probe the relationship between fiber and yarn structure, we first need to understand yarn structure. Yarns may be either filament or staple yarns. Filament yarns are made using continuous fiber; staple yarns are made from staple fiber. Industrial yarns of synthetic fibers, such as those used to reinforce rubber, in which mechanical properties are usually paramount, are almost exclusively flat filament yarns—yarns consisting of untextured, continuous fibers. Some filament yarns are textured and twisted, such as those used in sweaters and other textile applications. All yarns made using cotton and wool are staple yarns. Texturing is not required, since these fibers come with a natural texture that gives them bulk. On the other hand, synthetic staple fiber yarns are textured to give them bulk. All staple yarns are twisted to keep the filaments from separating.

The twist level in industrial filament yarns is kept to a minimum. In fact, filament yarns may simply be "intermingled" rather than twisted. Intermingling causes local entanglements that prevent fibers from separating when subjected to static charge. Twisting is kept to a minimum because, as we already know, both strength and modulus decrease with twist—strength decreases with $\cos^2 \theta$ and modulus decreases with $\cos^4 \theta$. In some special cases, however, filament yarns are twisted a considerable amount. These special cases involve applications in which the yarn strain-to-fail needs to be greater than that of the fibers. Recall that the strain-to-fail of a yarn increases with $1/\cos^2 \theta$. In twisted filament yarns the helix angle is $0°$ at the center of the yarn, where the fiber is straight, and maximum on the surface of the yarn. Consequently, fibers in the center of the filament experience more strain than do those on the yarn surface. Failure in a twisted-filament yarn initiates at the fibers in the center of the yarn.

Staple yarns differ from filament yarns in a number of significant ways. Their structure depends on how the yarn is made. Ring-spun yarns typically have all the fibers twisted together and arranged in concentric helices. It is the generic yarn we have used and will continue to use as the yarn model. Open-end-spun yarns have a core of twisted fibers and a sheath of wrapper fibers at various angles to the yarn axis. Air-jet-spun yarns have essentially a core of parallel fibers that are bound with wrapper fibers. The wrapper is another fiber that is oriented chiefly in the circumferential direction. The structures of ring, air-jet, and open-end yarns are shown in Figure 15.2. As exemplified in ring-spun yarns, fibers in staple yarns are given twist in processing to impart integrity to the yarn. Without twist, fibers simply slide past one another.

Another important distinction between filament and staple yarns is the role of migration. We have presented a simplified view of the path of a fiber in a yarn. We are now prepared to refine this view. The path of a single fiber does not follow a constant helix angle; rather, each fiber migrates from the center of the yarn to the surface of the yarn and back several times. When the yarn is tensioned, the section of the fiber in the center of the yarn experiences the maximum strain, and the stress is transmitted along the fiber, causing the yarns to tighten up and densify. Hearle sums it up: "Twist and fiber migration in spun yarns are not merely secondary factors as they are in continuous filament yarns; they are the only reason why an assembly of short fibers holds together as a yarn" (Hearle et al., 1969). In open-end and air-jet-spun yarns the effects are similar; however, some of the wrappers on the exterior may also provide significant lateral pressure on the internal fibers as the yarn is extended.

There are no fiber ends in a filament yarn. Since fiber ends cannot bear tensile stresses, filament yarns are stronger than their equivalent staple yarns, over and above the considerations given to twist. Fibers in a staple yarn transmit stresses from fiber to fiber via shear forces acting on the surface of the fiber. The greater the specific surface area or the smaller the fiber, the better. The higher the coefficient of friction, the better. Tensile stresses build up from

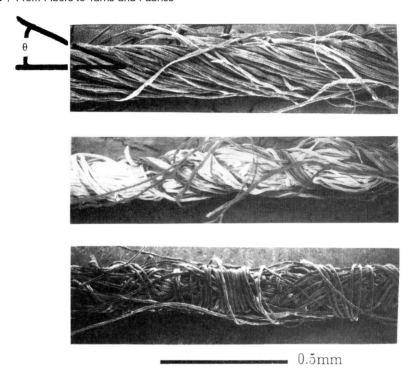

0.5mm

FIGURE 15.2 Structure of staple fiber yarns. (M. L. Realff, M. Seo, M. C. Boyce, P. Schwartz, and S. Backer, *Textile Research Journal* 61 (1991), 517–530.)

zero at the end of a fiber to roughly a constant level along the center portion of the fiber, as described in Chapter 4. Let us consider the transfer of stress from one fiber to another in a twisted yarn of staple fibers. If we for the moment treat the transfer of stress from one fiber to the next as a continuum, as is the case in fiber imbedded in a matrix, then we can sketch the stresses as a function of position along the length of a staple fiber, as shown in Figure 15.3. Shear stresses are a maximum at the ends of the fibers and reach zero when shear stresses are no longer required to transfer stresses to the fiber. Similarly, tensile stress is a constant maximum value in the center section of the fiber (assuming the fiber is more than twice its critical length, as defined in Chapter 4) and falls to zero at the fiber ends, where tensile stresses are precluded. Consider now the response of a highly twisted staple yarn to a tensile stress. The fibers transfer stress from one another via frictional forces, which are proportional to the normal forces. As the applied strain increases, the axial as well as the lateral forces increase. Because of migration, the twisted yarn densifies, perhaps from a low of 55 vol % to a high of 75 vol %. Failure is initiated generally by breakage of a fiber due to a combination of high stress and poor local mechanical properties. When the first fiber fails, the stress in the neighboring fibers increases. It takes only a relatively short axial distance, length l in Figure 15.3, from the failure site for the broken fiber to resume its

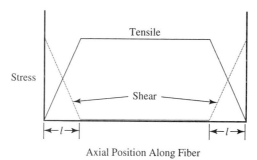

FIGURE 15.3   Tensile and shear stresses in a fiber in a twisted yarn.

fair share of the load. On the other hand, a broken fiber in a filament yarn with no or low twist will not be able to carry any load over its entire length. Tensile failure in filament yarns thus tends to be catastrophic.

These failure mechanisms are similar to those characteristic of fiber-reinforced composite materials. Models based on rigorous mechanical treatments of composite materials have been developed, but such analysis of staple fibers is lacking, chiefly because the situation is much more complex. While on a macroscopic scale a yarn is a one-dimensional object, the micromechanics of yarn failure require a two- or three-dimensional treatment. In addition, the structure and properties of fibers and yarns are highly irregular. Analytical treatments require insertion of properties into a model and often leave little room for statistical variations. The interested student is referred to the book by Hearle, Grosberg, and Backer listed in the reference section at the end of the chapter.

The most outstanding feature observed on data from experimental testing of staple yarns is that yarn tenacity increases with twist to a certain level and decreases thereafter, as shown in Figure 15.4. The effect is readily understood on the basis of the concepts already presented. A maximum in a curve such as this suggests two competing mechanisms. Increasing the coupling between the fibers causes the strength to increase. Twist increases the fiber-to-fiber friction. On the other hand, twist increases the helix angle, which reduces the axial load-carrying ability of each fiber. In the limit of zero friction, the tenacity is zero. In the limit of infinite twist, all the fibers are oriented circumferentially, and the axial strength is essentially zero. Somewhere between the limits is a maximum.

Let us consider the effect of fiber characteristics on yarn properties. Fiber straightness or crimp affects yarn density and the coupling between fibers. Fiber chemistry influences fiber-to-fiber friction. Fiber denier and cross-sectional shape influence surface area-to-volume ratio and, hence, fiber-to-fiber coupling. In addition, high denier fibers make poor low denier yarns. Long fibers give rise to yarns with fewer defects and fewer weak areas. Yarn structure also impacts yarn properties. How much migration exists? Is the yarn a ring- or core-spun yarn? What is the density of the yarn? The finishing step is also critical to staple yarn properties. Even a small change in fiber-to-fiber

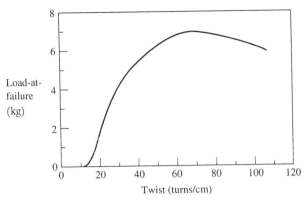

FIGURE 15.4   Effect of twist on the tenacity of cotton yarn. (A. E. Oxley, *Journal of the Textile Institute (Trans.)* 13 (1922), 54.)

friction can change both the surface chemical and mechanical properties of the yarn significantly. An adhesive finish transforms a yarn into a stiff fibrous composite.

15.2 Fabric Structure and Properties

Figure 15.1 shows schematically the relationship between yarn properties and fabric properties, and we now understand a little about the relationship between fiber and yarn properties. Let us begin our discussion of fabric properties with a brief description of woven and knit structures. Only the most basic aspects of woven and knit structures are covered—enough so that you can understand the elementary features of woven and knit fabric properties and deformation. Subsequently, we will describe nonwoven materials. Nonwoven structures are unique:

- Fabric is made directly from fibers.
- Fibers are anchored at the bonding points.

These two facts give nonwoven fabrics a host of unusual properties.

15.2.1 Structure of Woven Fabrics

Woven materials are formed by interlacing or intermeshing yarns. Let us consider only the simplest woven structure, the plain weave, shown in Figure 15.5. The direction of weaving is called the warp direction, and the normal direction is called the weft or fill. The path of a yarn through the fabric is modeled by the work of Peirce, reproduced in Figure 15.6. The warp yarns alternatively go over and under the fill. The angle that the warp and fill yarns assume between contact points is determined by the geometry of the fabric. When the fabric

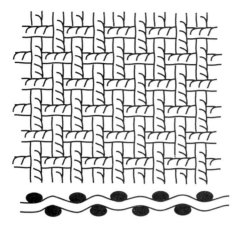

FIGURE 15.5 Structure of plain weave.

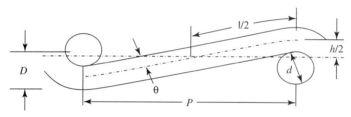

FIGURE 15.6 Region of plain weave. (F. T. Peirce, *Textile Research Journal* 17 (1947), 123.)

is subjected to a tensile stress, say along the warp direction, the straightening out of the yarns produces normal forces. The deformation produces

- Strain in the fabric.
- Flattening of the yarns, the extent depending on the transverse properties of the yarn.
- A reduction of the angle of the warp yarn between contact points.
- An increase in the angle of the fill yarn between contact points.

At sufficiently high levels of stress, warp yarns will begin to fail. Increased stress due to a yarn failure may be concentrated over only a local region, or, alternatively, broken yarns may be scattered over a wide area, depending on yarn interactions in the structure.

The modulus of the fabric increases with strain, since crimp is removed from the warp and weft yarns and the angle of the warp yarn between contact points decreases. Yarns may slide past one another unless the fill yarns become so closely spaced that they jam, as illustrated in Figure 15.7. Fabric construction, in this case the type of weave, the tensions used, and so on, contributes to the properties of the fabric. If we conduct a Gedanken, or thought,

FIGURE 15.7   Jamming in plain weave.

experiment, and maintain the fabric construction constant, we can begin to see the impact of yarn properties on fabric properties:

• The greater the yarn modulus, the greater the fabric modulus.
• The greater the yarn strength, the greater the fabric strength.
• The greater the yarn elongation, the greater the fabric elongation.

Hence, there is usually (not always) a direct translation of yarn mechanical properties into fabric mechanical properties, given a standard fabric structure. There are numerous subtle developments that we have not discussed that make fabric mechanics both interesting and challenging.

15.2.2 Structure of Knit Fabrics

Knitting involves interlooping of yarns. In weft knitting the needles drag new loops of yarn though old loops, as shown in Figure 15.8a. In warp knitting bars wrap new loops of yarn around the needles, as shown in Figure 15.8b. Since weft knitting accounts for the bulk of the knitting industry, we will focus on weft knit structures, although the concepts apply equally well to warp knit materials.

Consider the path of a single yarn through a jersey knit fabric, as shown in Figure 15.9. The yarn loops around the other yarns, so that the path of the yarn is rather tortuous. If you were to pull the ends of the yarn, then it would not require much force to strain the yarn. Hence, the modulus of knit fabrics is low, the strain knits can accommodate is high, and the elastic recovery is good. This stems largely from the fact that yarns pull on loops of other yarns, and yarns can flex quite a bit without inducing permanent deformation. The argument is similar to that used to explain the low modulus of rubbers, in which the applied tensile forces act to straighten out coiled molecules. Because knit structures have inherent stretch, they are usually made using yarns that are capable of considerable stretch and recovery. High-modulus, low-elongation fibers and yarns are typically not used in knit structures. Knit fabrics have good drape and conform well. When made using fine denier fibers, knit fabrics are soft and feel good against the skin. Hence, jersey knits are the fabric used in T-shirts, polo shirts, underwear, socks, and so on. Because of their open and flexible structure, they are also used for panty hose.

From this description it is evident how the properties of the knit fabric are derived from those of the yarn and the component fibers.

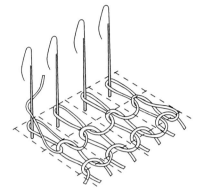

(a) Weft knitting

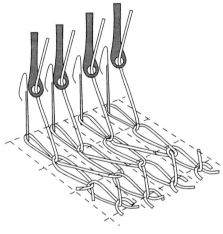

(b) Warp knitting

FIGURE 15.8 Weft and warp knit fabric formation. (Reprinted from D. J. Spencer, *Knitting Technology*, 2nd ed. (1989), 39, with kind permission from Pergamon Press Ltd, Headington Hill Hall, Oxford OX3 0BW, UK.)

15.2.3 Structure of Nonwoven Fabrics

Nonwoven fabrics are made directly from fibers. An intermediate yarn formation step is unnecessary. There are many ways to form a nonwoven fabric; however, we will mention only a few here. Nonwoven manufacturing consists of two steps: fiber laydown and bonding. Fiber laydown can be achieved by processing existing fibers, such as wet processing of wood pulp fibers to give paper, or carding of crimped synthetic or natural fibers. On the other hand, most of the new nonwoven processes convert synthetic polymer in the form of chip or pellet to fibers and lat them onto a web in one continuous process. Spunbonding is an example, illustrated in Figure 15.10. Polymer is melted in an

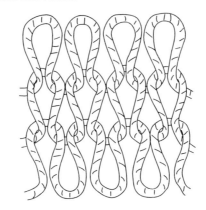

FIGURE 15.9   Path of yarn in jersey structure.

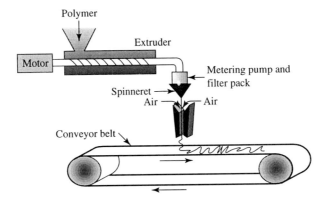

FIGURE 15.10   Schematic of spunbonding process.

extruder and delivered to a linear spinneret, whereupon the falling extrudate is attenuated using air flowing along the direction of the fiber axis. Emerging from the bottom of the air tubes or slots, the untextured fibers are laid directly on a conveyor belt. At this stage the web is formed, but the web lacks integrity. It must be bonded.

Bonding may be achieved using a number of different techniques: solvent, adhesive, melt, physical entanglement of fibers, or a combination. The industry is moving in the direction of entanglement bonding and thermal bonding. Entanglement bonding may be achieved using mechanical or hydraulic needling. In mechanical needling, barbed steel needles are repeatedly pushed through a nonwoven fabric, causing the staple fibers that get caught in the barbs to travel through the fabric, entangling the web. Most felts are made this way. Thinner fabrics consisting of staple fibers may be needled using arrays of high-velocity water jets that impinge on the face of the fabric, driving fiber ends through the fabric and entangling the fibers. Alternatively, thermal bonding is a fusion bonding technique that may use patterned calender rolls, as shown in Figure 15.11, or ultrasonic techniques when the fabric is heavyweight, or may be

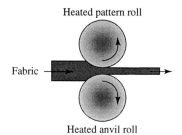

Heated pattern roll

Fabric

Heated anvil roll

FIGURE 15.11   Schematic of calendaring process.

achieved by through-air bonding when bicomponent fibers with a low melting temperature component are used. In the calendar bonding technique, the web, say the spunbond web described in the previous paragraph, is fed into the nip defined by a smooth (anvil) roll and an engraved (pattern) roll. Both rolls are heated, and the combined action of heat and pressure causes the fibers to fuse under the raised portions of the engraved roll. Using a combination of these techniques, a wide variety of nonwoven materials may be manufactured with a range of useful properties at extraordinary speeds—orders of magnitude faster than woven or knit materials can be produced. Examples include diaper liner, sterile wrap, and geotextiles.

The properties of nonwoven fabrics are determined by the structure of the fabric and the constituent fibers. Wetting and absorbency characteristics, for example, depend on the surface energy of the fibers and the pore size distribution within the fabric. The mechanical properties of the fabric depend on those of the fibers, the orientation distribution of the fibers in the fabric, and the nature of the bonding. We illustrate how the bending behavior of a fabric depends on the fabric construction and the properties of the fibrous components in the following section. The bending behavior of fabrics is a key property in a variety of applications. In fact, poor drape or bending stiffness has prevented nonwoven materials from penetrating the apparel market.

15.3 Bending Behavior of Fabrics

Among the chief factors that differentiate film from fabric are their drape characteristics. Films are not able to bend in two directions simultaneously, as illustrated in Figure 15.12a. Many textile materials are required to show good drape. In order to be able to drape well, a planar material needs to be able to bend easily and undergo shear at small stresses. Low bending stiffness is the dominant factor. This can be demonstrated using a series of tests, as indicated in Figures 15.12a–c (Cusick, 1962). In Figure 15.12, a 30 cm diameter round planar sample is clamped symmetrically between two discs 18 cm in diameter. Drape is a measure of how the sample hangs from the discs. No drape is when the sample stays planar. Films bend in only one plane. Knit fabrics bend readily in two planes simultaneously. Rigid fabrics, such as many nonwovens,

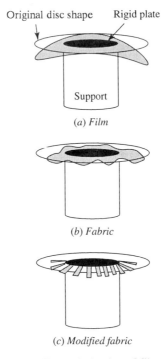

Original disc shape   Rigid plate

Support

(a) Film

(b) Fabric

(c) Modified fabric

FIGURE 15.12   Drape behavior of film and fabric.

show drape characteristics more like those of a film than a knit. When the rigid sample is modified by removing radial wedges, as shown in Figure 15.12c, the effect of shear is eliminated. Clearly, if a sample still fails to bend, then the drape is zero. The difference between cases b and c gives an indication of the effect of shear.

Kawabata has developed a system that facilitates design of fabrics based on sets of characteristics required of fibers, yarns, and fabrics. The system utilizes more than a decade of studies in which a great quantity of subjective (sensory) and analytical data were correlated. Perhaps you are or will become familiar with the series of test equipment that Kawabata developed for analysis of textile materials.

15.3.1 Drape of Knit Fabrics

It is intuitively obvious why knit samples have good drape:

- The yarn path is tortuous.
- The fabric structure is open.
- Loops of yarns can redistribute to accommodate strain.
- The yarns are typically of low bending modulus.

These properties account for both low shear modulus and low bending rigidity and hence good drape of knit fabrics.

15.3.2 Drape of Woven Fabrics

Let us consider the drape of woven fabrics. The drape characteristic of a woven fabric depends on a number of factors.

- Consider first the amount of twist in the yarn. The higher the yarn twist, the stiffer the yarn in bending. This effect can be understood in terms of the effect of radius on the bending stiffness of a rod. Recall from Chapter 9 that the bending stiffness of a round fiber is given as

$$EI = E\pi R^4/4$$

where

E = fiber tensile modulus,
I = moment of inertia, and
R = fiber radius.

A bundle of n untwisted fibers, each of radius R, has a bending stiffness given by

$$EI = nE\pi R^4/4$$

Each fiber is free to act independently; however, when the bundle is twisted highly, the fibers act like one large rod, radius $R\sqrt{n}$. Consequently, the bending stiffness of the rod is given by

$$EI = E\pi(R\sqrt{n})^4/4 = n^2 E\pi R^4/4$$

The bending stiffness of the twisted bundle is n times that of the untwisted fibers, neglecting the increase in radius and the decrease in modulus (both of which are relatively small effects) that result from the twisting.

- Another factor that affects the drape of a woven fabric, again assuming a balanced simple weave, is the tightness of the weave. Woven structures can be made that have near zero air permeability, indicating the yarns are packed tightly. The tight packing precludes substantial motion of fibers or yarns to relieve stresses. Hence, shear resistance is high.

In sum, the drape of a woven fabric can be low or moderate, depending on the fabric structure, the structure of the yarns, and the properties of the fibers. Moderate drape is desired in some textile applications, such as shirts; low drape is required in others, such as men's suit coats.

15.3.3 Drape of Nonwoven Fabrics

Like woven and knit fabrics, drape in nonwoven fabrics depends chiefly on bending behavior, but also on shear behavior. We will consider just two nonwoven fabrics, a hydraulically needled fabric and a calendar-bonded spunbond fabric. Using these two examples, we can show the important concepts and treatments. In addition, these two fabrics perhaps represent extremes in bending behavior.

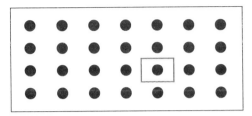

FIGURE 15.13   Pattern-bonded nonwoven fabric.

Fibers in nonwoven fabrics are not packed as tightly as they are in twisted yarns. There is always considerable space between fibers, and the fabric density is generally much lower than that of a yarn. Hence, in the absence of bonding, fibers are essentially free to move relative to one another. This gives the material low shear modulus and low bending modulus, shear being largely controlled by the fabric construction and bending by the number of fibers per unit area, the fiber diameter, and the fiber modulus. Because of their low density, however, nonwoven fabrics are generally thick materials. Recall from Chapter 8, p. 159, that the bending stiffness of a slab per unit width is given by

$$EI = Et^3/12$$

where t is the slab thickness. Hence, small increases in thickness have large effects on the bending stiffness of a slab.

Bonding can change the drape properties of a fabric by orders of magnitude. Mechanical or hydraulic needling creates a bonded structure by entangling fibers; however, the fibers retain a large portion of their independence. Hence, a needled fabric typically has good drape, unless the fabric is unusually thick. When the fabric is deformed, fibers slide past one another, tightening up the entanglements like knots. When the stress is removed, the fabric recovers little from the deformation. Most of the deformation is permanent. That is, the fabrics "bag," much like wool after extended use, especially when wet or damp.

Consider now the calendar-bonded, spunbond fabric shown in Figure 15.13. The discrete consolidated regions are the bond points, where the fabric has been transformed essentially into a biaxially oriented film. Typically the bonded area represents 5 to 20 percent of the total area. The continuous region is the unbonded fabric.

The bending behavior of the fabric has been analyzed by treating the material as a two-region material. The bonded regions are considered to be a film or thin slab. We already understand bending of a slab. The bending rigidity of the unbonded regions is calculated in two ways, following Freeston and Platt (Freeston and Platt, 1965):

1. The fibers are treated as completely free to move. Hence, the fibers in the unbonded regions behave as a collection of independent fibers. The thickness of the unbonded region is not important in this model, but fiber diameter is.

(a) Unit cell   (b) Unit cell for Takayanagi

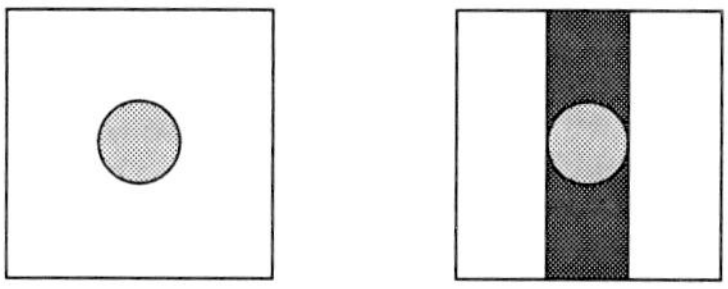

FIGURE 15.14   Division of nonwoven fabric into unit cell.

2. The fibers in the unbonded region are treated as not free to move. Hence, the fabric acts like a low-density slab. Thickness is important in the model and fiber diameter is not.

The two preceding models give results that differ by 4 to 6 orders of magnitude. From this difference the importance of relative fiber motion to bending stiffness and drape can be readily understood.

To proceed with our model for the bending stiffness of point-bonded fabrics, we need to assess what fraction of the fibers behave as free to move. The balance are not free to move. We calculate this by noting that fibers anchored by closeby bond regions are not free to move. Consequently, we use probability theory to assess the portion of fibers that enter adjacent bond regions. These fibers are not free to move. This gives us a value for the bending stiffness of the unbonded regions (Warner and Patel, 1993).

We now have the bending stiffness of bonded and unbonded regions. We need to combine them to obtain a value for the bending stiffness of the fabric. We note that the fabric has a repeat pattern, as indicated in Figure 15.13. We can define a unit cell, much like we did for a crystal in Chapter 3. The properties of the unit cell are the same as those of the entire material. Shown in Figure 15.14 is the unit cell of the nonwoven fabric. Takayanagi has shown how to combine a two-phase micellar model such as this to facilitate calculation of mechanical properties. Substituting into the appropriate Takayanagi expression, we obtain an expression for bending stiffness of a calender-bonded spunbond fabric (Takayanagi et al., 1966):

$$(EI)_{\text{fabric}} = (EI)_1 V_1 + [(EI)_2 V_2 + (EI)_b V_b]^{-1}$$

where

V's are volume fractions,

b refers to bonded areas,

1 represents the unbonded area shaded in Figure 15.14b, and

2 represents the unshaded unbonded area in Figure 15.14b.

For a spunbond web with no preferred fiber orientation and round fibers,

$$(EI)_b = E_f t_b^3 / 12$$

and

$$(EI)_u = \{0.125 X I_f [3 + 1/(1 + \nu)] + 0.031(1 - X) A_f t^2 u\} N E_f$$

where

X = weight fraction of fibers not free to move,

I_f = moment of inertia of the fiber = $\pi R^4 / 4$,

ν = Poisson's ratio of the fiber,

A_f = cross-sectional area of the fiber = πR^2,

t_u = thickness of the unbonded region,

N = number of fibers per unit cell, and

E_f = modulus of the fiber.

The model has been found to give results that agree very well with experimental values of bending stiffness. While I do not expect that you understand all the details of the model from such a quick treatment, I do expect you to appreciate that the expression shows that bending stiffness depends completely on basic fiber properties and fabric construction.

To complete our analysis of drape in this nonwoven fabric, we need to consider shear resistance. Shear modulus of a nonwoven fabric will also depend on the relative freedom of fibers to move independently. We can return to the calculations made on this and the previous page to get a handle on shear resistance. While the details of such an analysis have not yet been worked out, we can note that the shear resistance increases with relative bond area.

15.4 Summary

Yarns are usually twisted assemblies of fibers. Yarns may be filament yarns or staple yarns, depending on the fibers from which they are constructed. The structure of a staple yarn depends on how it was formed. Fiber migration is caused by the need to relieve stresses in fibers on the outside of the yarn during formation. While migration occurs in both filament and staple yarns, it is critical to the behavior of staple yarns. Without migration in ring-spun yarns, stresses would not be transmitted from fiber to fiber. The mechanical properties of a yarn, such as an ideally helically twisted yarn, depend on the helix angle θ. Modulus decreases with $\cos^4 \theta$, elongation increases with $1/\cos^2 \theta$, and strength decreases with $\cos^2 \theta$.

Knit and woven fabrics are made from yarns. Knit fabrics are typically able to sustain high deformations and show good recovery. This is largely because the yarns' path through the undeformed, loosely knit structure is tortuous. Adjacent loops readily deform, permitting the easy achievement of high fabric deformation, and, in addition, knits are usually made using stretchy yarns, which also promotes fabric deformation. Woven fabrics may be much denser than knit fabrics, suggesting that yarn motion may be hindered. Initial properties, such as fabric modulus along a principal direction, can be calculated from the properties of the yarn corrected for the angle the yarn takes through the fabric. Ultimate properties also depend on fabric structure and yarn properties.

Nonwoven fabrics are composed of fibers themselves, not yarns. The fibers may be continuous or staple fibers, textured or flat. Bonding may be achieved using a variety of techniques, including thermal and entanglement techniques. The properties of the fabric may be calculated from the fabric structure, the properties of the fibers, and the properties of the bonds. We considered drape as an example. Drape depends chiefly on bending stiffness, but also on fabric shear. We explained the concepts and presented mathematical expressions that allow the calculation of drape or bending stiffness from fundamental fiber and fabric properties in point-bonded nonwovens.

References

S. Backer and D. R. Petterson. *Textile Research Journal* 30 (1960), 704.

G. E. Cusick, Ph.D. thesis, University of Manchester, 1962.

W. D. Freeston and M. M. Platt. "Mechanics of Elastics Performance of Textile Materials"; Part VI, "Bending Rigidity of Nonwoven Fabric." *Textile Research Journal* 35 (1965), 48.

J. W. S. Hearle, P. Grosberg, and S. Backer. *Structural Mechanics of Fibers, Yarns and Fabrics*. New York: Wiley-Interscience, 1969.

S. Kawabata. *The Standardization and Analysis of Hand Evaluation*, 2nd ed. Textile Machinery Society of Japan, July 1980.

P. R. Lord. *The Economics, Science, and Technology of Yarn Production*. Manchester, England: The Textile Institute, 1981.

N. Pan. *Textile Research Journal* 63 (1993), 336, 504, and 565.

D. J. Spencer. *Knitting Technology*, 2nd ed. New York: Pergamon Press, 1989.

M. Takayanagi, K. Imada, and T. Kajiyama. *Journal of Polymer Science: Part C* 15 (1966), 263.

E. Vaughn. "Principles of Woven, Knitted, and Nonwoven Fabric Formation." Clemson University, 1992.

S. B. Warner and S. V. Patel. "Predicting Nonwoven Fabric Bending Behavior." Atlanta: TAPPI, 20 April 1993.

Problems

(1) A woven fabric is produced so that it shows good drape. Describe the fibers, yarns, and fabric structure required to achieve the result. If the fabric is extrusion-coated with a thin (8 μm thick) layer of polyethylene, how will the drape be affected?

(2) How might Kawabata develop a woven fabric with good drape?

(3) Would you select filament or staple yarns for use in an automobile tire? Answer the question first by stating the requirements of such a reinforcing yarn.

(4) If the main requirement of a fabric is to have good cover, what materials and what structure would you select to achieve the result?

(5) Compare the density of real yarns with the theoretical density of an assembly of parallel close-packed cylinders. Explain the difference.

(6) An untwisted yarn has a tenacity of 4.7 g/d. Sketch the change in tenacity with increasing twist. On the same set of axes, plot the effect of helix angle only.

(7) A yarn is used in an environment in which the fibers become coated with a slick lubricant. Why might a filament yarn be preferred to a staple yarn in this application?

(8) Kevlar® fibers have low elongation-to-break. If the fibers fail at 2% strain and use in an automobile tire requires 4% strain, how can the fibers be used?

(9) An elastomer is stretched several times its original length. While under tension, a water-soluble coating is applied to the stretched elastomeric filament yarn and the solvent is allowed to evaporate. The coating has become a rigid structure that holds the fiber stretched out. Describe what occurs in the composite when it is used as part of a legband in a diaper.

(10) Describe qualitatively and quantitatively (if possible) how the following changes affect the bending stiffness of a thermally bonded nonwoven fabric:

(a) from round to ribbon fiber cross-sectional shape.

(b) from PET to PP fiber.

(c) making the fabric thinner and, hence, denser.

(d) adding a finish to reduce fiber–fiber friction.

(11) Calculate the relative bond area of the fabric shown here.

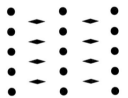

(12) Describe an ideal parachute fabric.

(13) An elastic band presses down on a nonwoven fabric, which presses against the skin, as in an elasticized cuff. What fabric properties determine how the downward elastic force is spread over the width of the cuff?

16

Fluid Interactions

Another important property of aggregates of fibers is fluid absorption, the interactions of fluids with fiber surfaces and the spaces between fibers. Paper towels or diapers, for example, need to absorb and distribute large quantities of aqueous fluids. Polypropylene shop towels are used to pick up oils. We are on firm ground to begin, since we have aquired a fundamental understanding of bulk absorption of moisture into hygroscopic fibers. This chapter focuses on water as the fluid, but the underlying principles apply to all fluids. The interaction of water with fibers becomes important when the fluid water is present. Water is present when it is added directly to the fabric or when the relative humidity is at or near 100 percent. When a dry natural fiber is dampened with water, absorption and fluid interactions occur simultaneously. Although we are concerned with the motion of moisture in both cases, the processes are different. Fluid motion is driven by gravimetric and capillary forces. As we already showed, absorption into the bulk of the fiber occurs by diffusion in a concentration gradient.

When a paper towel is said to be absorbent, it means that the towel can pick up a large mass of fluid, usually a water-based spill. The amount of fluid taken up by an assembly of fibers is generally much greater than the amount taken up by the sum of the individual fibers. That is, most of the fluid is trapped between the fibers and fiber absorption is negligible. Consider, for example, the wood pulp in a diaper. The assembly may trap ten times its weight in fluid, yet we know that semicrystalline cellulose fibers themselves absorb less than 25 percent under even ideal conditions. The majority of the fluid is trapped between fibers whose surfaces appeal to the water. Let us first examine the interactions between fiber surfaces and fluids.

16.1 Surface Energy, Tension, and Wetting

Consider a tiny drop of fluid, no larger than the head of a pin, that has been placed on a horizontal solid surface, as shown in Figure 16.1. Every surface has excess energy associated with it. In a primitive sense the excess energy may be envisioned as being due to unsatisfied bonds at the surface of the fluid. For example, a water molecule on the surface of a drop does not have nearest neighbors completely surrounding it as do those molecules in the bulk of the drop. The polarity of the molecule on the surface is not completely neutralized by neighbors. Hence, the molecules right on the surface are in a higher energy state than those in the bulk. We use the symbol γ to represent surface energy. The units of γ are energy/unit area. In recent history the units for surface energy were ergs/cm^2, where an erg $= 10^{-7}$ J. In fluids surface energy, γ_{L-V}, gives rise to a force per unit length, which is a surface tension. Surface tension has the same units as surface energy. The old units are dynes/cm $=$ (ergs/cm)/cm $=$ ergs/cm^2. The new units for surface tension are milli-Newtons/m $=$ mN/m. Surface tension and surface energy are quantitatively the same in fluids.

In the case of the drops shown in Figure 16.1, there are three surface energies:

- γ_{S-V} = surface energy between the solid and the vapor;
- γ_{S-L} = surface energy between the solid and the drop;
- γ_{L-V} = surface energy between the liquid and the vapor.

The angle that the drop forms between the horizontal surface and the fluid–air interface, θ, is determined by the surface energies just described. We can establish the relationships among these variables by conducting a force balance across the horizontal surface at the point where the surface of the drop meets the solid flat surface, shown by the dot on the left side of Figure 16.1:

$$\gamma_{S-V} = \gamma_{S-L} + \gamma_{L-V}\cos\theta$$

This expression is known as the Young–Dupre equation. The left side of the equation is the force to the left, and the right side of the equation is the sum of the resolved forces acting to the right.

When $\theta \leq 90°$, then the drop is said to wet the surface. Perfect wetting corresponds to a contact angle of 0°; this is also referred to as a self-spreading

FIGURE 16.1 Drops of fluids on a flat surface.

TABLE 16.1   Contact Angles of Various Polymers

Polymer	Contact Angle			
	Ethanol	*Toluene*	*Ethylene Glycol*	*Water*
Polypropylene	47		74	86
PET	26	56	61	75
Nylon	18	57	57	71

Source: B. Miller in M. J. Schick, ed., *Surface Characteristics of Fibers and Textiles*, Part II, New York: Marcel Dekker, 1977.

TABLE 16.2   Water Wetting and Work of Adhesion of Various Fibers

Fiber	Work of Adhesion (mN/m)
Scoured cotton	148
Oxide glass	138
Carbon	128
Nylon	106
Polyester	85
Polypropylene	77
Teflon	56
Wool	46

Source: B. Miller in M. J. Schick, ed., *Surface Characteristics of Fibers and Textiles*, Part II, New York: Marcel Dekker, 1977.

drop. When $\theta < 90°$, the energy of the drop is less when it has a large area of contact with the solid surface than when the drop is riding high on the surface, touching the solid surface minimally. When $\theta < 90°$, the drop prefers the solid surface to the vapor. When $\theta > 90°$, wetting does not occur. The fluid and the surface are not capable of secondary bonding. When water is the fluid, a nonwetting surface is generally not capable of hydrogen bonding. You may correctly anticipate that hydrophilic ("water-loving") surfaces are characteristic of hygroscopic fibers. Fluid and surface generally need similar solubility parameters for good wetting. Contact angles for fluids on polymers are compiled in Table 16.1. The total attraction per unit area between fluid and solid is called the work of adhesion, W_{S-L}. It can be defined thermodynamically:

$$W_{S-L} = \gamma_{L-V} + \gamma_{S-V} - \gamma_{S-L}$$

Combining this expression with the Young–Dupre equation yields

$$W_{S-L} = \gamma_{L-V}(1 + \cos\theta)$$

By examining the work of adhesion rather than the contact angle, scientists can compare the wetting behavior of various fluids on solid surfaces. Table 16.2 provides data on work of adhesion of various fibers exposed to water. The surface tension of water is 73.05 mN/m (dynes/cm). Consider, for example,

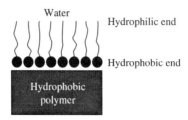

Water

Hydrophilic end

Hydrophobic end

Hydrophobic polymer

FIGURE 16.2 Schematic of surfactant action on hydrophobic polymer to make it water wettable.

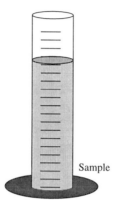

Sample

FIGURE 16.3 Hydrohead experiment for fabrics.

whether oil or water will wet polypropylene fabric. Oil is a linear hydrocarbon with a structure similar to that of low-molecular-weight polyethylene. Hence, it is capable of van der Waals bonding only. Water is a polar molecule, capable of strong hydrogen bonds. PP is wet by oil and not wet by water. PP, PE, and other olefins can be treated with molecules that have an olefinic group on one end and a polar group on the other end. These surfactants, which can be designed for any fluid–solid pair, allow water to wet polyolefins, as shown schematically in Figure 16.2. Surfactants can be added to the melt and bloom to the surface or padded onto the polymer after it is shaped.

When a fluid does not wet a polymer, the fluid rides high on the surface, such as mercury on the floor or water on a polytetrafluoroethylene (PTFE), or Teflon®, pan. This property can be used to advantage in fabrics. Water droplets will not pass through small holes in a hydrophobic or nonwetting fabric unless they are forced to do so; however, air and other vapors may pass through the small holes. This is the basis of most water-repellent yet breathable fabrics, such as Goretex®. Goretex® is essentially a microporous PTFE film laminated to a fabric substrate (D. J. Bucheck, *Chemtech* (March 1991), 142).

Hydrohead is one technique used to quantify water repellency. A column of water is placed on top of a fabric, as shown in Figure 16.3. The height

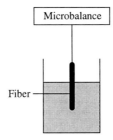

FIGURE 16.4 Schematic of Wilhelmy technique used to assess fiber wetting.

of the column prior to breakthrough is a measure of the resistance to water penetration.

The expression for contact angle assumes a perfectly flat surface. We have not considered the effect of surface curvature, even though we described the hydrohead test for fabrics. The effect of curved surfaces depends on the wetting characteristics and the nature of the curved surface. When a surface is wet by a fluid, making the surface rough on a microscale further decreases the contact angle, improving the apparent wetting character. On the other hand, when the surface is nonwetting and air is trapped between the rough spots, such as might occur in a fibrous substrate, then the apparent contact angle increases, making the surface seem even more hydrophobic. This technique is used by ducks. The contact angle of water on feather material is only 95°, but a feather is composed of many fine fibers, or hairs, roughly 8 μm in diameter. As a result, the water appears to have a contact angle of 150°, so water rolls off the duck's back!

How, then, do we measure the contact angle on fibers? Sometimes it is possible to use a model surface, such as a film. Other times it is necessary to use the Wilhelmy technique, shown in Figure 16.4. In this technique a fiber is lowered into the fluid of interest. The mass of the fiber is continuously monitored as the fiber is lowered into the fluid. The wetting force, F_w, is balanced by the gravimetric force, F_g:

$$F_w = L\gamma\cos\theta = F_g = Mg$$

where

L is the contact length of the fluid with the fiber and air,
M is the mass,
g is acceleration due to gravity (980 cm/sec^2), and
the other symbols have their usual meaning.

For a fiber with a round cross-section, radius R,

$$L = 2\pi R$$

Hence,

$$\cos\theta = Mg/2\pi R\gamma$$

The Wilhelmy technique was originally used for plates. Single fiber work requires an excellent experimentalist using sensitive equipment. Only then can good data be obtained. (I have intentionally ignored some of the details associated with the Wilhelmy technique, such as the fact that a weight is often used to keep the fiber straight, requiring calculation of the buoyancy forces as the fiber and weight are immersed.)

16.2 Pressure Across a Curved Interface

In this section we are interested in developing a quantitative expression that begins to allow us to determine whether a system of capillaries will absorb fluid, given a contact angle θ. The capillaries are not typically holes in the fibers, but rather are the interstices between fibers, such as present in woven, knit, or nonwoven fabrics. The denser the fabric, the smaller the capillary cross-section and the smaller the total capillary volume.

Consider a system of parallel fibers. The minimum volume fraction of pores in a system of equal-sized round fibers can be calculated by considering the fibers to be hexagonally close-packed, as shown in Figure 16.5. The triangle described by the locus of points lying on the lines connecting the centers of three mutually contacting fibers describes a unit cell. The unshaded area within the triangle divided by the area of the triangle gives the volume fraction of pores:

$$V_{\text{shaded}} = 3(1/6)\pi R^2 = \pi R^2/2$$

and

$$V_{\text{triangle}} = (R)R\sqrt{3} = R^2\sqrt{3}$$

Thus,

$$V_{f \text{ unshaded}} = [V_{\text{triangle}} - V_{\text{shaded}}]/V_{\text{triangle}}$$
$$= [\sqrt{3} - \pi/2]/\sqrt{3} = 9\%$$

The volume fraction of void in a parallel array of closely packed round fibers is 9 percent. Conversely, the maximum volume fraction of fiber in yarn or uniaxial composite of undeformed round fiber is 91 percent. For ease of calculation we will approximate the intersticies between fibers as cylindrical capillaries.

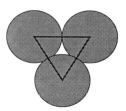

FIGURE 16.5   Unit cell of close-packed fibers.

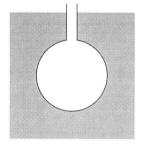

FIGURE 16.6 Blowing a bubble in a fluid.

16.2.1 Capillarity

Surface energy causes a pressure difference to develop across a curved surface. This is the fundamental law of capillarity, alternatively called the LaPlace equation. To understand the origin of the pressure gradient across a curved surface, insert a capillary or straw into a liquid and blow a bubble, as shown in Figure 16.6. The resistance to expansion of the bubble is due to the increase in interfacial surface area with corresponding surface energy, γ:

$$\Delta P \, dV = \gamma \, dA$$

where

ΔP = pressure drop across the bubble interface,

dV = differential change in volume of the bubble, and

dA = differential change in interfacial surface area of the bubble.

Note that both sides of the preceding equation are work terms. The increase in differential volume of a sphere is readily calculated, since it is the differential surface area:

$$dV = 4\pi R^2 \, dR$$

and

$$dA = 8\pi R \, dR$$

Rearranging the work expression in terms of ΔP and substituting the values for differential volume and surface area,

$$\Delta P = \gamma (dA/dV) = \gamma (8\pi R \, dR / 4\pi R^2 \, dR) = 2\gamma / R$$

As previewed, the expression states that the pressure drop is caused by the curved interface. As $R \to \infty$ (a flat interface), the pressure drop goes to 0. ΔP increases with decreasing bubble diameter, showing that bubble nucleation in a fluid is very difficult.

The ΔP causes liquids to rise in capillaries in which the fluid wets the capillary surface, as shown in Figure 16.7. The rise of the fluid in the column is limited by the hydrostatic pressure of the fluid column:

$$\Delta P = \rho g h$$

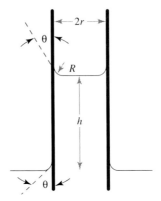

FIGURE 16.7 Fluid rise in a capillary,

where

ρ = density of the fluid,

g = acceleration due to gravity, and

h = height of the fluid in the capillary.

Equating the two expressions for pressure drop,

$$\rho g h = 2\gamma/R$$

Solving for γ,

$$\gamma = R\rho g h/2$$

But we note that using simple geometry the curvature of the fluid, R, is related to the capillary radius, r, in small capillaries:

$$R = r/\cos\theta$$

Hence,

$$\gamma = r\rho g h/2\cos\theta$$

This expression shows one way that the elusive γ can be measured. It is also apparent from the mathematics that as the capillary radius decreases, the fluid, perhaps water, will rise to a higher level in the capillary. Although the height of water in the smaller capillary is greater, less water can be absorbed by an equal number of small capillaries, since $V \propto \pi r^2 h$. In the next section we will show how difficult it is to remove fluid from a capillary.

16.2.2 Capillary Suction

Consider the effect of centrifugal force on the height of fluid in a capillary. In this calculation we assume an ideal orientation of the capillaries, with the capillary axis oriented radially in a centrifugal force field, as shown in Figure 16.8. With this geometry we do not need to resolve centrifugal or capillary

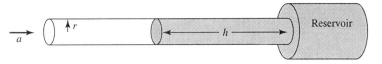

FIGURE 16.8 Centrifugation of fluid in a cylindrical capillary.

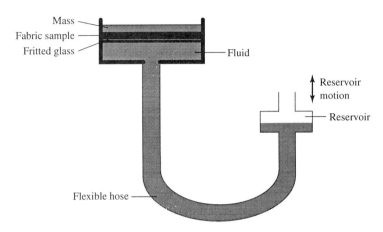

FIGURE 16.9 Capillary suction test.

forces. Gravitational forces are negligible. The solution is the best case for fluid removal—an upper bound calculation. The force acting to remove the fluid from the capillary, mass times acceleration, is balanced by the holding forces related to surface energy:

$$\pi r^2 h \rho a = 2\pi r \gamma \cos \theta$$

where

a = acceleration due to centrifuging, and

the other symbols have their usual meaning.

We can solve this expression for h to give the height of the column of fluid in the capillary during centrifugation:

$$h = 2\gamma \cos \theta / r \rho a$$

The height varies inversely with a. When the fluid is water, the maximum height the fluid can rise in the capillary is, of course,

$$h_{max} = 2\gamma / r a$$

When the centrifuge is stopped, the fluid will backfill the capillary unless the fluid reservoir loses contact with the capillary.

An experimental procedure that is designed to provide information on pore size and pore size distribution is the capillary suction test. In this test a saturated fabric is drained using negative hydrohead or hydrostatic tension, as shown in Figure 16.9. In the test the reservoir is lowered and raised, and the

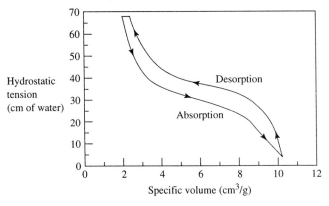

FIGURE 16.10 Data from capillary suction test on rayon fabric. (A. Burgeni and C. Kapur, *Textile Research Journal* 37 (1967), 356.)

volume of water drained from the sample is measured at all times. Large capillaries do not hold water as tenaciously as do small capillaries, so they drain first. Hence, information can be learned not only on capillary volume, as shown in Figure 16.10, but also on capillary size distribution. From these data, the capillary size distribution can be calculated using the previous expression, realizing that fluid drains or fills pores larger than a critical size:

$$r_c = 2\gamma \cos\theta / pgh$$

and the volume of a pore is that of a cylinder. Companies concerned with absorption by fibrous structures, such as those in the diaper and absorbent paper industries, use this or a similar test routinely.

16.3 Flow in Capillaries

We have learned to what height a fluid will spontaneously rise in a capillary. The fluid may be envisioned as being sucked up the capillary, driven by the favorable fluid–solid interaction. The extent of fluid rise depends on the capillary radius, the fluid, and surface interactions. When the capillary is on its side, the same phenomenon occurs—the fluid moves into the capillary—but the flow is not stopped by the rising height and mass of fluid. Rather, a flow rate develops. How much fluid can be moved from one point to another in a capillary system?

The Lucas–Washburn equation provides the information we need. It can be expressed in many ways, and we have selected a convenient form that is also one of the most common forms. As illustrated in Figure 16.11, the Lucas–Washburn equation gives an expression for the velocity v of a fluid with viscosity η in a tube with length l and radius r:

$$v = dx/dt = r\gamma \cos\theta / 4\eta l$$

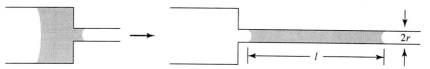

FIGURE 16.11 Horizontal flow in a capillary.

The equation states that the fluid velocity increases with the diameter of the tube, the solid–fluid surface energy, and $\cos\theta$, and decreases with the fluid viscosity and length of the tube. Recall that the volume of fluid that can move down a tube varies with the diameter squared. Thus, decreasing the diameter of a capillary has a huge effect on the volume of fluid that can be moved by a capillary.

An example where this sort of behavior is important is in a diaper. The urine is delivered to a local area, and the ideal diaper moves the fluid away from the body, away from the delivery point, and contains the urine even under the stresses imposed by body weight.

Other expressions that characterize the flow of fluid through porous media have been developed and may be useful. One well-known expression is Darcy's law, which states that the rate of flow though a porous media is proportional to the pressure drop across the media. In terms of a vertically downward flow, the pressure drop is related to the piezometric head, Δh:

$$Q = KA\Delta h/L,$$

where

Q = the volumetric flow,
A = the cross-sectional area of the porous media,
K = a constant inversely proportional to viscosity, and
$\Delta h/L$ is related to the pressure drop.

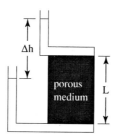

Darcy's law can be extended to multidimensional flow analysis:

$$Q = -K(\nabla \cdot \varphi) = -K(\partial\varphi/\partial x + \partial\varphi/\partial y + \partial\varphi/\partial z)$$

where φ is a pressure field.

FIGURE 16.12   Moving material from a flat interface to a spherical interface.

Darcy's law was developed in the mid 1800's to describe water flow through sand filters, but has been used extensively in other systems (Bear, 1972), including fibrous assemblies (Miller, 1992).

16.4 The Kelvin Equation

The Kelvin equation provides us with a mathematical relationship that shows, among other things, that a fluid in small droplets evaporates faster than the same material in large droplets. We anticipate this, knowing that curved surfaces have excess energy and a pressure drop across them.

A relatively primitive way to derive the Kelvin equation is to consider the work done in transferring 1 mole of material from a flat interface to a spherically curved interface across the vapor, as shown in Figure 16.12. The work done in moving the material is PV work:

$$\text{Work} = RT\ln(p/p_o)$$

where

p = vapor pressure over the curved surface, and

p_o = vapor pressure over a flat interface.

The work in forming new surface is

$$\text{Work} = \gamma\, dA$$

For a sphere of radius r and volume V,

$$\gamma\, dA = \gamma 8\pi r\, dr = \gamma 8\pi r(V/4\pi r^2)$$

Since the two work terms are equal

$$RT\ln(p/p_o) = \gamma 2V/r \Rightarrow \ln(p/p_o) = [V\gamma/RT](2/r)$$

Another form of the Kelvin equation is often used in composite or absorbent technology to calculate the level of pressure required to reduce porosity to a small value. This form of the Kelvin equation can be applied to situations where the matrix has not completely wet-out fibers at the fiber–fiber intersections, as shown in Figure 16.13. The appropriate form of the Kelvin equa-

FIGURE 16.13 Voids at fiber intersections.

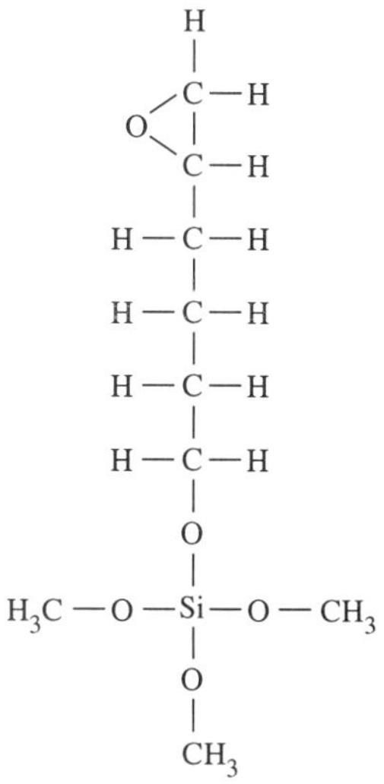

$$
\begin{array}{c}
\text{H} \\
|\\
\underset{}{\overset{}{\text{O}}}\diagdown\text{C}-\text{H}\\
|\\
\text{C}-\text{H}\\
|\\
\text{H}-\text{C}-\text{H}\\
|\\
\text{H}-\text{C}-\text{H}\\
|\\
\text{H}-\text{C}-\text{H}\\
|\\
\text{H}-\text{C}-\text{H}\\
|\\
\text{O}\\
|\\
\text{H}_3\text{C}-\text{O}-\text{Si}-\text{O}-\text{CH}_3\\
|\\
\text{O}\\
|\\
\text{CH}_3
\end{array}
$$

FIGURE 16.14 Schematic of silane coupling agent used to bond oxide glass to epoxy. The top portion includes an epoxy group that opens and covalently bonds with the polymer. The bottom port: n contains a reactive silicon alkoxide that bonds with silica.

tion is

$$P = 4\gamma \cos\theta / r$$

where

P = pressure required to cause the void to shrink to radius of curvature r.

Squeeze casting is a common technique for forming metal–matrix composites in cases where the liquid metal matrix hesitates to wet the fiber, such as when molten aluminum is used with alumina, silicon carbide, or carbon fibers. It should be evident, however, that if the resin does not at least partially wet the fiber, then a good bond at the interface and, hence, a good composite cannot be formed.

When matrix does not wet fiber, coupling agents may be used to improve wetting and hence bonding between fiber and matrix. Without the coupling agent, fiber and matrix do not approach sufficiently closely to form secondary or primary bonds. A coupling agent is a material that is generally applied to the fiber. One end of the coupling agent forms a primary bond to the fiber. The other end is compatible with the matrix. It either bonds to the matrix or simply entangles with the matrix. Consider, for example, bonding of oxide glass fibers to epoxy. The silane coupling agent used forms a primary bond with the silica (half ionic, half covalent) and has a structure similar to that of the polymer, as shown in Figure 16.14. One end is a silicon alkoxide that readily reacts with silica, and the other end is a hydrocarbon that entangles or bonds with the polymer. Coupling agents have been developed for a variety of fiber–matrix systems. Earlier in the text we discussed surfactants. An important difference between surfactants and coupling agents is that the latter form strong bonds between the surfaces and are thus permanent. Surfactants are usually padded on surfaces, often can wash off, and are at best semipermanent.

16.5 The Campbell Effect

Suppose you are asked to test two nonwoven fabrics that have been wet-formed from staple fibers. One is made using staple oxide glass fibers and the other is made using wood pulp fibers. Your first test is on the dry fabrics and then you test them wet. Is there any difference in strength between wet and dry fabrics? The answer is yes. The fabric consisting of wood pulp fibers is strongly hydrogen-bonded at fiber crossover points. It is and has the properties of a sheet of paper. Hydrogen bonding in the wood pulp fabric is disrupted when water is added, so the wet-strength is low. The glass fiber fabric may show the opposite effect. The fibers have only weak bonds between them, as dry silica fibers do, so the dry fibrous assembly is weak. On the other hand, in the wet state water binds the fibers according to the strength of the water–fiber bond. To break the water–fiber bond requires formation of additional surface area of water. This is the Campbell effect. There are numerous applications where fibers are processed wet or with oils and the Campbell effect, plus effects due to the viscosity of the fluid, binds fibers that seem otherwise unbound. Although often small, the effect can be appreciable, especially in the absence of major forces.

Consider the importance of the Campbell effect in papermaking. To make paper, wood pulp fibers are added to vast amounts of water, making a slurry of fiber in water. The mixture is allowed to drain through a screen, or wire as it is called in the paper industry, leaving a nonwoven web of fibers completely saturated with water. The fabric at this point has little, but some, strength. As the water evaporates, however, the surface tension of the water pulls the fibers together, as shown in Figure 16.15, with force

$$F = 2\gamma L$$

where

L = fiber contact length.

When they are sufficiently close, the fibers hydrogen bond together, giving the paper strength. Without the Campbell effect, papermaking as we know it would not be possible.

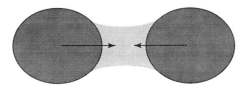

FIGURE 16.15 Surface tension drawing fibers together. *Note*: Wood pulp fibers are not round, but flat.

16.6 Summary

We have concentrated on the behavior of fluids in contact with fibrous substrates. The fluid with which we have been most concerned is water, largely because aqueous-based solutions are the target applications of many knit, woven, and nonwoven products, such as diapers, bath towels, and paper towels, and water is used in wet processing of staple fibers. We have introduced not only many major concepts, but also the associated mathematics, which facilitates quantitative use of the concepts.

The contact angle of a fluid against a solid surface gives information regarding the ability of the fluid to wet and spontaneously spread across the solid surface. By maintaining large contact angles, textile scientists are able to make breathable yet water-repellent fabrics. Providing a small contact angle ensures that a fluid will rise in a capillary and fill small pores in a fabric. Water binds tenaciously to small pores or capillaries, yet large capillaries carry large volumes of water under the appropriate driving force. Capillary suction is one test that may be conducted on a material to learn about its pore size distribution.

Fluids, including water, have a surface tension that may produce significant effects in materials. A single drop of water, for example, can make it difficult to separate two glass slides. Water is also the force that pulls wood pulp fibers together and allows formation of paper. Finally, it is essentially impossible to eliminate all the void from a fluid–fibrous composite unless the contact angle is zero. Pressure will reduce the void size, but some void will always persist.

References

J. Bear. *Dynamics of Fluid in Porous Media*. New York: Dover Publications, 1972.
F. Brouchard-Wyart, J.-M. diMeglio, and D. Quere. *Journal of Physics, France* 51 (1990), 293.
P. Chatterjee, ed. *Absorbency*. New York: Elsevier, 1985.
B. W. Cherry. *Polymer Surfaces*. Melbourne, Australia: Cambridge University Press, 1981.
J. A. Cornie, Y.-M. Chiang, D. R. Uhlmann, A. Mortenson, and J. M. Collins. "Processing of metallic and ceramic matrix composites." *Ceramic Bulletin* 65 (1986), 293.
W. D. Kingery, H. K. Bowen, and D. R. Uhlmann. *Introduction to Ceramics*, 2nd ed. New York: Wiley-Interscience, 1976.

B. Miller in M. J. Schick, ed. *Surface Characteristics of Fibers and Textiles*, Part II. New York: Marcel Dekker, 1977.

B. Miller. *Textile Research Journal* 62 (1992), 151–161.

J. Swanson (at the Institute of Paper Science and Technology, Atlanta, GA), Fiber Science short course, 25 October 1985, Neenah, WI.

Problems

(1) Sketch the profile of a small oil droplet resting on the surface of cellophane, wood, PP, PE, and PET. Comment on contact angles.

(2) Is it more difficult to drink milk from small-diameter straws than from large-diameter straws? Note that modern straws are PE. Justify your answer.

(3) Two fabrics are made using PP fibers. One fabric contains large fibers that trap a great deal of air between the fibers. The other fabric contains small fibers that are closely spaced. Comment on the absorbency of the two fabrics, each treated with surfactant to make it water wettable.

(4) We know from extensive studies that water wets both cotton and wool. PET is the fiber most widely used to replace cotton and wool. Explain whether you expect PET to be wet by water. If you expect $\theta > 90°$, then what may be done to improve the situation? (Hint: Consider the Young–Dupre equation.)

(5) How much pressure is required to reduce the size of the voids to 1 μm radius in a composite of alumina fibers in an aluminum matrix, given the surface energy of molten Al is 500 J/m^2 and the contact angle with Al_2O_3 is 110°?

(6) Explain why it is so difficult to obtain information regarding wetting behavior from fibers or fibrous assemblies. Describe an experiment for learning about θ of fibers in fluids.

(7) Polypropylene nonwoven fabrics allow a great deal of air to pass through under low pressure, but the same fabric can be used to contain water when shaped into a cup. Why does the water not pass through?

(8) Your assignment is to make a better diaper. It is well established that one of the major roles of the material in a diaper, wood pulp fiber, is to move fluid from the point of discharge and absorb as much fluid as possible. Discuss what might be done to

(a) get the fluid to move faster and farther.

(b) bind the fluid more tightly.

(c) provide for greater fluid capacity.

(9) What is a good polymer to use to absorb oil from spills on the ocean? What is a good form of the polymer to facilitate high capacity? Is fluid viscosity a consideration?

(10) Two different liquids are placed on an assembly of fibers. Which way will the drops move? Why?

(11) Describe a fibrous assembly that is ideal for oil recovery in the calm ocean. Do you anticipate that PP will be better than cotton? (H.-M. Choi, H.-J. Kwon, and J. P. Moreau. *Textile Research Journal* 63 (1993), 211.)

(12) Cotton towels are stiff and boardy after drying on the clothesline. What if they were freeze-dried (frozen wet and the ice sublimed)?

(13) A 10 g sample of benzene is dispersed as droplets each with radius 1.0 μm. The surface tension of benzene is 0.028 N/m and its density is 0.88 g/cm^3. How much work is required to achieve the dispersal?

(14) At room temperature the surface tension of ethanol in contact with its vapor is 0.022 N/m and its density is 0.78 g/cm^3. If a straw with internal diameter 0.20 cm is inserted into the ethanol, how far up the straw will the ethanol rise, given the contact angle is **(a)** 30° and **(b)** 95°? How much pressure will it require to lower the level of the fluid to match that of the flat interface?

(15) Rain X® is an automotive product that is applied to car windshields. It makes the rain roll off the windshield, eliminating the need for windshield wipers when driving at high speed. How does Rain X® work?

Appendix I

This text uses a variety of units, largely because this simulates real life. You should be familiar with a number of systems of units and be able to convert into a system meaningful to you.

SI Prefixes

giga, g	10^9	milli, m	10^{-3}
mega, m	10^6	micro, μ	10^{-6}
kilo, k	10^3	nano, n	10^{-9}

Length

$1 \text{ Å} = 10^{-10} \text{ m} = 0.1 \text{ nm}$

$1 \text{ m} = 10^{-6} \text{ m} = 10^{-4} \text{ cm}$

$1 \text{ cm} = 10^{-2} \text{ m}$

Mass

$454 \text{ g}_m = 1 \text{ lb}_m$

Force

$1 \text{ N} = 1 \text{ kg}_m\text{-m/sec}^2$

$1 \text{ N} = 0.2248 \text{ lb}_f$

Stress, Modulus, Pressure

$1 \text{ Pa} = 1 \text{ N/m}^2$

$1 \text{ Pa} = 1.019 \times 10^{-7} \text{ kg}_f/\text{mm}^2$

$1 \text{ Pa} = 1.450 \times 10^{-4} \text{ lb}_f/\text{in}^2$

$1 \text{ MPa} = \text{ksi} \times 6.895$

Specific Stress or Modulus

Breaking length, km $= 9 \times g_f/\text{denier}$

Viscosity	**Temperature**
1 Pa-sec = 10 poise	$°C = K - 273$
	$°C = (5/9)(°F - 32)$

Current	**Frequency**	**Capacitance**	**Energy**
1 A = 1 C/sec	Hz (cycle/sec) = sec^{-1}	farad, F = C/V	1 J = 1 W-sec
1 A = 1 V/Ω			1 J = 1 N-m
			1 J = 1 V/C
			1 J = 0.239 cal
			1 J = 0.738 ft-lb$_f$

Other Conversion Factors

$$1 \text{ N/tex} = 102 \times g_f/\text{tex} = 11.3 \text{ g/denier}$$
$$1 \text{ MPa} = 88.26 \times (g_f/\text{denier}) \times \text{density}$$
$$1 \text{ ksi} = 12.80 \times (g_f/\text{denier}) \times \text{density}$$
$$\text{tex} = 9 \times \text{denier}$$
$$2.54 \text{ cm} = 1 \text{ in}$$

Constants

Acceleration to gravity on earth, g	9.807 m/sec^2
Avogadro's number, N_A	6.023×10^{23}/mole
Atomic mass unit, amu	1.661×10^{-24} g
Electric permittivity of vacuum, ε_o	8.854×10^{-12} C/(V-m)
Gas constant, R	8.134 J/(mole-K)
	or 1.987 cal/(mole-K)
Boltzmann's constant, k	13.81×10^{-24} J/K
Planck's constant, h	6.626×10^{-34} J-sec
Speed of light in vacuum, c	2.998×10^8 m/sec

Appendix II

PERIODIC TABLE OF ELEMENTS

Table of Selected Radioactive Isotopes

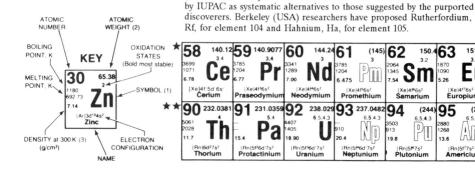

†The names and symbols of elements 104–106 are those recommended by IUPAC as systematic alternatives to those suggested by the purported discoverers. Berkeley (USA) researchers have proposed Rutherfordium, Rf, for element 104 and Hahnium, Ha, for element 105.

(1) Based upon carbon-12. ( ) indicates most stable or best known isotope.
(2) Entries marked with asterisks refer to the gaseous state at 273 K and 1 atm and are given in units of g/l.

Selected Radioactive Isotopes

Naturally occurring radioactive isotopes are designated by a mass number in blue (although some are also manufactured) Letter m indicates an isomer of another isotope of the same mass number Half-lives follow in parentheses, where s, min, h, d, and y stand respectively for seconds, minutes, hours, days, and years. The table includes mainly the longer-lived radioactive isotopes; many others have been prepared. Isotopes known to be radioactive but with half-lives exceeding 10^{15} y have not been included. Symbols describing the principal mode (or modes) of decay are as follows (these processes are generally accompanied by gamma radiation)

α   alpha particle emission
β⁻   beta particle (electron) emission
β⁺   position emission
EC   orbital electron capture
IT   isomeric transition from upper to lower isomeric state
SF   spontaneous fission

VIII

2	4.00260
4.215	
0.95	
0.1787*	**He**
	$1s^2$
	Helium

IIIB	IVB	VB	VIB	VIIB	

5 10.81	6 12.011	7 14.0067	8 15.9994	9 18.998403	10 20.179
3	±4.2	±3,5,4,2	−2	−1	
4275 / 2300	4470* / 4100*	77 35 / 63 14	90 18 / 50 35	84 95 / 53 48	27 096 / 24 553
2.34 **B**	2.62 **C**	1.251* **N**	1.429* **O**	1.696* **F**	0.901* **Ne**
$1s^22s^22p^1$	$1s^22s^22p^2$	$1s^22s^22p^3$	$1s^22s^22p^4$	$1s^22s^22p^5$	$1s^22s^22p^6$
Boron	Carbon	Nitrogen	Oxygen	Fluorine	Neon

13 26.98154	14 28.0855	15 30.97376	16 32.06	17 35.453	18 39.948
	4	±3,5,4	±2,4,6	±1	
2793 / 933.25	3540 / 1685	550 / 317.30	717.75 / 388.36	239 1 / 172 16	87 30 / 83 81
2.70 **Al**	2.33 **Si**	1.82 **P**	2.07 **S**	3.17* **Cl**	1.784* **Ar**
[Ne]$3s^2p^1$	[Ne]$3s^2p^2$	[Ne]$3s^2p^3$	[Ne]$3s^2p^4$	[Ne]$3s^2p^5$	[Ne]$3s^2p^6$
Aluminum	Silicon	Phosphorus	Sulfur	Chlorine	Argon

IB	IIB					

29 63.546	30 65.38	31 69.72	32 72.59	33 74.9216	34 78.96	35 79.904	36 83.80
2,1					±2,4,6	±1,5	
1836 / 357 6	1180 / 692.73	2478 / 302.90	3107 / 1210.4	876 (subl.) / .081 (28 atm)	958 / 494	332 25 / 265 90	119 80 / 115.78
.96 **Cu**	7.14 **Zn**	5.91 **Ga**	5.32 **Ge**	5.72 **As**	4.80 **Se**	3.12 **Br**	3.74* **Kr**
[Ar]$3d^{10}4s^1$	[Ar]$3d^{10}4s^2$	[Ar]$3d^{10}4s^2p^1$	[Ar]$3d^{10}4s^2p^2$	[Ar]$3d^{10}4s^2p^3$	[Ar]$3d^{10}4s^2p^4$	[Ar]$3d^{10}4s^2p^5$	[Ar]$3d^{10}4s^2p^6$
Copper	Zinc	Gallium	Germanium	Arsenic	Selenium	Bromine	Krypton

47 107.868	48 112.41	49 114.82	50 118.69	51 121.75	52 127.60	53 126.9045	54 131.30
1			4,2	±3,5	±2,4,6	±1,5,7	
1436 / 234	1040 / 594.18	2346 / 429.76	2876 / 505.06	1860 / 904	1261 / 722.65	458 4 / 386 7	165 03 / 161 36
0.5 **Ag**	8.65 **Cd**	7.31 **In**	7.30 **Sn**	6.68 **Sb**	6.24 **Te**	4.92 **I**	5.89* **Xe**
[Kr]$4d^{10}5s^1$	[Kr]$4d^{10}5s^2$	[Kr]$4d^{10}5s^2p^1$	[Kr]$4d^{10}5s^2p^2$	[Kr]$4d^{10}5s^2p^3$	[Kr]$4d^{10}5s^2p^4$	[Kr]$4d^{10}5s^2p^5$	[Kr]$4d^{10}5s^2p^6$
Silver	Cadmium	Indium	Tin	Antimony	Tellurium	Iodine	Xenon

79 196.9665	80 200.59	81 204.37	82 207.2	83 208.9804	84 (209)	85 (210)	86 (222)
3,1	2,1	3,1	4,2	3,5	4,2	±1,3,5,7	
30 / 337 58	630 / 234 28	1746 / 577	2023 / 600 6	1837 / 544 52	1235 / 527	610 / 575	211 / 202
9.3 **Au**	13.53 **Hg**	11.85 **Tl**	11.4 **Pb**	9.8 **Bi**	9.4 **Po**	**At**	9.91* **Rn**
[Xe]$4f^{14}5d^{10}6s^1$	[Xe]$4f^{14}5d^{10}6s^2$	[Xe]$4f^{14}5d^{10}6s^2p^1$	[Xe]$4f^{14}5d^{10}6s^2p^2$	[Xe]$4f^{14}5d^{10}6s^2p^3$	[Xe]$4f^{14}5d^{10}6s^2p^4$	[Xe]$4f^{14}5d^{10}6s^2p^5$	[Xe]$4f^{14}5d^{10}6s^2p^6$
Gold	Mercury	Thallium	Lead	Bismuth	Polonium	Astatine	Radon

* Estimated Values

The A & B subgroup designations, applicable to elements in rows 4, 5, 6, and 7, are those recommended by the International Union of Pure and Applied Chemistry. It should be noted that some authors and organizations use the opposite convention in distinguishing these subgroups.

65 158.9254	66 162.50	67 164.9304	68 167.26	69 168.9342	70 173.04	71 174.967
3,4	3	3	3	3,2	3,2	3
96 / 30	2835 / 1682	2968 / 1743	3136 / 1795	2220 / 1818	1467 / 1097	3668 / 1936
27 **Tb**	8.54 **Dy**	8.80 **Ho**	9.05 **Er**	9.33 **Tm**	6.98 **Yb**	9.84 **Lu**
[Xe]$4f^96s^2$	[Xe]$4f^{10}6s^2$	[Xe]$4f^{11}6s^2$	[Xe]$4f^{12}6s^2$	[Xe]$4f^{13}6s^2$	[Xe]$4f^{14}6s^2$	[Xe]$4f^{14}5d^16s^2$
Terbium	Dysprosium	Holmium	Erbium	Thulium	Ytterbium	Lutetium

97 (247)	98 (251)	99 (252)	100 (257)	101 (258)	102 (259)	103 (260)
4,3						
	900					
Bk	**Cf**	**Es**	**Fm**	**Md**	**No**	**Lr**
[Rn]$5f^97s^2$	[Rn]$5f^{10}7s^2$	[Rn]$5f^{11}7s^2$	[Rn]$5f^{12}7s^2$	[Rn]$5f^{13}7s^2$	[Rn]$5f^{14}7s^2$	[Rn]$5f^{14}6d^17s^2$
Berkelium	Californium	Einsteinium	Fermium	Mendelevium	Nobelium	Lawrencium

TABLE OF PERIODIC PROPERTIES

Percent Ionic Character of a Single Chemical Bond

Difference in electronegativity	0.1	0.2	0.3	0.4	0.5	0.6	0.7	0.8	0.9	1.0	1.1	1.2	1.3	1.4	1.5	1.6	1.7	1.8	1.9	2.0
Percent ionic character %	0.5	1	2	4	6	9	12	15	19	22	26	30	34	39	43	47	51	55	59	63

GROUP IA

H
0.32 | 2.20
0.79 | 0.44936
14.4 | 0.058.68
13.598 | —
14.304 | 0.001815

DATA CONCERNING THE MORE STABLE ELEMENTARY (SUBATOMIC) PARTICLES

	Neutron	Proton	Electron*	Neutrino*	Photon
Symbol	n	p	e⁻ (e⁻)	ν	γ
Rest mass (kg)	1.67495×10^{-27}	1.67265×10^{-27}	9.1095×10^{-31}	~0	0
Relative atomic mass ($^{12}C = 12$)	1.008665	1.007276	5.48580×10⁻⁴	~0	0
Charge (C)	0	1.60219×10^{-19}	-1.60219×10^{-19}	0	0
Radius (m)	8×10^{-16}	8×10^{-16}	$<1 \times 10^{-16}$	~0	0
Spin quantum number	1/2	1/2	1/2	1/2	1
Magnetic Moment†	$-1.913 \, \mu_N$	$2.793 \, \mu_N$	$1.001 \, \mu_B$	0	0

*The positron (e⁺) has properties similar to those of the (negative) electron or beta particle except that its charge has opposite sign (+). The antineutrino ($\bar{\nu}$) has properties similar to those of the neutrino except that its spin (or rotation) is opposite in relation to its direction of propagation.

An antineutrino accompanies release of an electron in radioactive β (particle) decay, whereas a neutrino accompanies the release of a positron in β^+ decay. †μ_B = Bohr magneton and μ_N = Nuclear magneton.

IIA

Li
1.23 | 0.98
2.05 | 145.920
13.10 | 3.00
5.392 | 0.108
3.6 | 0.847

Be
0.90 | 1.57
1.40 | 292.40
5.0 | 12.20
9.322 | 0.313
1.82 | 2.00

Na
1.54 | 0.93
2.23 | 96.960
23.7 | 2.598
5.139 | 0.210
1.23 | 1.41

Mg
1.36 | 1.31
1.72 | 127.40
13.97 | 8.954
7.646 | 0.226
1.02 | 1.56

IIIA / IVA / VA / VIA / VIIA / VIIIA

K
2.03 | 0.82
2.77 | 79.870
45.46 | 2.334
4.341 | 0.139
0.75 | 1.024

Ca
1.74 | 1.00
2.09 | 153.60
29.9 | 8.540
6.113 | 0.298
0.63 | 2.00*

Sc
1.44 | 1.36
2.09 | 314.20
15.0 | 14.10
6.54 | 0.0177
0.6 | 0.158

Ti
1.32 | 1.54
1.92 | 421.00
10.64 | 15.450
6.82 | 0.0234
0.52 | 0.219

V
1.22 | 1.63
1.92 | 0.452
8.78 | 20.90
6.74 | 0.0489
0.49 | 0.307

Cr
1.18 | 1.66
1.85 | 344.30
7.23 | 16.90
6.766 | 0.0774
0.45 | 0.937

Mn
1.17 | 1.55
1.79 | 226.0
7.39 | 12.050
7.435 | 0.00695
0.48 | 0.0782

Fe
1.17 | 1.83
1.72 | 349.60
7.1 | 13.80
7.870 | 0.0993
0.44 | 0.802

Co
1.16 | 1.88
1.67 | 376.50
6.7 | 16.190
7.86 | 0.172
0.42 | 1.00

Ni
1.15 | 1.91
1.62 | 370.40
6.59 | 17.470
7.635 | 0.143
0.44 | 0.907

Rb
2.16 | 0.82
2.98 | 72.216
55.9 | 2.192
4.177 | 0.0779
0.363 | 0.582

Sr
1.91 | 0.95
2.45 | 144.0
33.7 | 8.30
5.695 | 0.0762
0.30 | 0.353

Y
1.62 | 1.22
2.27 | 363.0
19.8 | 11.40
6.38 | 0.0166
0.30 | 0.172

Zr
1.45 | 1.33
2.16 | 58.20
14.1 | 16.90
6.84 | 0.0236
0.27 | 0.227

Nb
1.34 | 1.6
2.08 | 682.0
10.87 | 26.40
6.88 | 0.0693
0.26 | 0.537

Mo
1.30 | 2.16
2.01 | 598.0
9.4 | 32.0
7.099 | 0.187
0.25 | 1.38

Tc
1.27 | 1.9
1.95 | 660.0
8.5 | 24.0
7.28 | 0.067
0.21 | 0.506

Ru
1.25 | 2.2
1.89 | 595.0
8.3 | 24.0
7.37 | 0.137
0.238 | 1.17

Rh
1.25 | 2.28
1.83 | 493.0
8.3 | 21.50
7.46 | 0.211
0.242 | 1.50

Pd
1.28 | 2.20
1.79 | 357.0
8.9 | 17.60
8.34 | 0.0950
0.24 | 0.718

Cs
2.35 | 0.79
3.34 | 67.740
71.07 | 2.092
3.894 | 0.0489
0.24 | 0.359

Ba
1.98 | 0.89
2.78 | 142.0
39.24 | 7.750
5.212 | 0.030
0.204 | 0.184*

La
1.69 | 1.10
2.74 | 414.0
20.73 | (20
5.58 | 0.0126
0.19 | 0.135

Hf
1.44 | 1.3
2.16 | 575.0
13.6 | 24.060
6.65 | 0.0312
0.14 | 0.230

Ta
1.34 | 1.5
2.09 | 743.0
10.90 | 31.60
7.89 | 0.189
0.14 | 0.575

W
1.30 | 2.36
2.02 | 824.0
9.53 | 35.40
7.98 | 0.189
0.13 | 1.74

Re
1.28 | 1.9
1.97 | 715.0
8.85 | 33.20
7.88 | 0.0542
0.13 | 0.479

Os
1.26 | 2.2
1.92 | 746.0
8.49 | 31.80
8.7 | 0.109
0.13 | 0.876

Ir
1.27 | 2.20
1.87 | 604.0
8.54 | 26.10
9.1 | 0.197
0.130 | 1.47

Pt
1.30 | 2.28
1.83 | 510.0
9.10 | 19.60
9.0 | 0.0966
0.13 | 0.716

Fr
— | 0.7
— | —
— | —
0.15* | —

Ra
— | 0.9
45.20 | —
5.279 | —
0.03 | —
0.186* | —

Ac
— | 1.1
22.54 | —
5.17 | —
— | 0.12*

Unq
— | —
— | —
— | —
— | 0.23*

Unp
— | —
— | —
— | —
— | 0.58*

Unh
— | —
— | —
— | —

†The names and symbols of elements 104–106 are those recommended by IUPAC as systematic alternatives to those suggested by the purported discoverers. Berkeley (USA) researchers have proposed Rutherfordium, Rf, for element 104 and Hahnium, Ha, for element 105. Dubna researchers, who also claim the discovery of the elements have proposed different names (and symbols

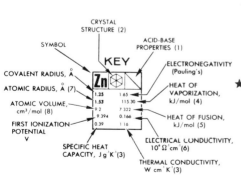

KEY

CRYSTAL STRUCTURE (2)
SYMBOL
ACID-BASE PROPERTIES (1)
Zn
COVALENT RADIUS, Å
ATOMIC RADIUS, Å (7)
1.25 | 1.65
1.53 | 115.30
9.2 | 7.322
9.394 | 0.166
0.39 | 1.16
ELECTRONEGATIVITY (Pauling's)
HEAT OF VAPORIZATION, kJ/mol (4)
ATOMIC VOLUME, cm³/mol (8)
FIRST IONIZATION POTENTIAL V
HEAT OF FUSION, kJ/mol (5)
SPECIFIC HEAT CAPACITY, J g⁻¹ K⁻¹ (3)
ELECTRICAL CONDUCTIVITY, $10^4 \, \Omega^{-1} \, cm^{-1}$ (6)
THERMAL CONDUCTIVITY, W cm⁻¹ K⁻¹ (3)

★ Lanthanides

Ce
1.65 | 1.12
2.70 | 414.0
20.67 | 5.460
5.54 | 0.0115
0.19 | 0.114

Pr
1.65 | 1.13
2.67 | 296.80
20.8 | 6.890
5.46 | 0.0148
0.19 | 0.125

Nd
1.64 | 1.14
2.64 | 273.0
20.6 | 7.140
5.53 | 0.0157
0.19 | 0.165

Pm
1.63 | 1.13
2.62 | —
22.39 | —
5.554 | —
— | 0.179*

Sm
1.62 | 1.17
2.59 | 166.40
19.95 | 8.630
5.64 | 0.00956
0.20 | 0.133

★★ Actinides

Th
1.65 | 1.3
— | 514.40
19.9 | 16.10
6.08 | 0.0653
0.12 | 0.540

Pa
— | 1.5
— | —
15.0 | 12.30
5.89 | 0.0529
— | 0.47*

U
1.42 | 1.38
— | 477.0
12.59 | 38.50
6.05 | 0.0380
0.12 | 0.276

Np
— | 1.36
— | —
11.62 | 5.190
6.19 | 0.00822
0.12 | 0.063

Pu
— | 1.28
— | 344.0
12.32 | 2.840
6.06 | 0.00666
0.13* | 0.0674

OF THE ELEMENTS

2.1	2.2	2.3	2.4	2.5	2.6	2.7	2.8	2.9	3.0	3.1	3.2
67	70	74	76	79	82	84	86	88	89	91	92

VIII

He
0.92
0.49 0.0845
24.587
5.193 0.00152

	IIIB	IVB	VB	VIB	VIIB	

B
0.82 2.04
1.17 489.70
4.6 50.20
8.298 10^{12}
102 0.270

C
0.77 2.55
0.91 355.80
4.58 —
11.260 0.00061
0.71 1.29

N
0.75 3.04
0.75 2.7928
17.3 0.3604
14.534 —
1.04 0.000259

O
0.73 3.44
0.65 3.4099
14.0 0.22259
13.618 —
0.92 0.000267

F
0.72 3.98
0.57 3.2698
17.1 0.2552
17.422 —
0.82 0.000279

Ne
0.71
0.51 1.7326
16.7 0.3317
21.564 —
0.904 0.000493

Al
1.18 1.61
1.82 293.40
10.0 10.790
5.986 0.377
0.90 2.37

Si
1.11 1.90
1.46 384.220
12.1 50.550
8.151 2.5×10^{-12}
0.71 1.48

P
1.06 2.19
1.23 12.129
17.0 0.657
10.486 10^{17}
0.77 0.00235

S
1.02 2.58
1.09 —
15.5 1.7175
10.360 5×10^{24}
0.71 0.00269

Cl
0.99 3.16
0.97 10.20
22.7 3.203
12.967 —
0.48 0.000089

Ar
0.98
0.88 6.447
28.5 1.188
15.759 —
0.520 0.0001772

IB	IIB	

Cu
1.17 1.90
1.57 300.30
7.1 13.050
7.726 0.596
0.38 4.01

Zn
1.25 1.65
1.53 115.30
9.2 7.322
9.394 0.166
0.39 1.16

Ga
1.26 1.81
1.81 258.70
11.8 36.940
5.999 0.0678
0.37 0.406

Ge
1.22 2.01
1.52 330.90
13.6 36.940
7.899 1.5×10^{8}
0.32 0.599

As
1.16 2.18
1.33 34.760
13.1 —
9.81 0.0345
0.33 0.500

Se
1.16 2.55
1.22 37.70
16.45 6.694
9.752 10^{12}
0.32 0.0204

Br
1.14 2.96
1.12 15.438
23.5 5.286
11.814 —
0.473 0.00122

Kr
1.12
1.03 9.029
38.9 1.638
13.999 —
0.248 0.0000941

Ag
1.34 1.93
1.75 250.580
10.3 11.30
7.576 0.630
0.235 4.29

Cd
1.48 1.69
1.71 99.570
13.1 6.192
8.993 0.138
0.23 0.968

In
1.44 1.78
2.00 231.50
15.7 3.263
5.786 0.116
0.23 0.816

Sn
1.41 1.96
1.72 295.80
16.3 7.029
7.344 0.0917
0.227 0.666

Sb
1.40 2.05
1.53 77.140
18.23 19.870
8.641 0.0288
0.21 0.243

Te
1.36 2.1
1.42 52.550
20.5 17.490
9.009 2×10^{4}
0.20 0.0235

I
1.33 2.66
1.32 20.752
25.74 7.824
10.451 8×10^{16}
0.214 0.00449

Xe
1.31
1.24 12.636
37.3 2.297
12.130 —
0.158 0.0000568

Au
1.34 2.54
1.79 334.40
10.2 12.550
9.225 0.452
0.128 3.17

Hg
1.49 2.00
1.76 59.229
14.82 2.295
10.437 0.0104
0.139 0.0834

Tl
1.48 2.04
2.08 164.10
17.2 4.142
6.108 0.0617
0.13 0.461

Pb
1.47 2.33
1.81 177.70
18.17 4.799
7.416 0.0481
0.13 0.353

Bi
1.46 2.02
1.63 104.80
21.3 11.30
7.289 0.00867
0.12 0.0787

Po
1.46 2.0
1.53 —
22.23 —
8.42 0.0219
0.20* —

At
(1.45) 2.2
1.43 —
— —
— —
0.017* —

Rn
—
1.34 16.40
50.5 2.890
10.748 —
0.09* 0.0000364

The A & B subgroup designations, applicable to elements in rows 4, 5, 6, and 7, are those recommended by the International Union of Pure and Applied Chemistry. It should be noted that some authors and organizations use the opposite convention in distinguishing these subgroups.

Eu
1.85 1.2
2.56 143.50
28.9 9.210
5.67 0.0112
0.18 0.139*

Gd
1.61 1.20
2.54 359.40
19.9 10.050
6.15 0.00736
0.23 0.106

Tb
1.59 1.2
2.51 330.90
19.2 10.80
5.86 0.00889
0.18 0.111

Dy
1.59 1.22
2.49 230.0
19.0 11.060
5.94 0.0108
0.17 0.107

Ho
1.58 1.23
2.47 241.0
18.7 12.20
6.018 0.0124
0.16 0.162

Er
1.57 1.24
2.45 261.0
18.4 19.90
6.101 0.0117
0.17 0.143

Tm
1.56 1.25
2.42 191.0
18.1 16.840
6.184 0.0150
0.16 0.168

Yb
1.74 1.1
2.40 128.90
24.79 7.660
6.254 0.0351
0.15 0.349

Lu
1.56 1.27
2.25 355.90
17.78 18.60
5.43 0.0185
0.15 0.164

Am
— 1.3
— —
17.86 14.40
5.993 0.022
0.11* 0.1*

Cm
— 1.3
— —
18.28 15.0
6.02 —
— 0.1*

Bk
— 1.3
— —
6.23 —
0.1*

Cf
— 1.3
— —
6.30 —
0.1*

Es
— 1.3
— —
6.42 —
0.1*

Fm
— 1.3
— —
6.50 —
0.1*

Md
— 1.3
— —
6.58 —
0.1*

No
— 1.3
— —
6.65 —
0.1*

Lr
— 1.3
— —
— —
0.1*

Index